# CPSS Power Electronics Series

This series comprises advanced textbooks, research monographs, professional books, and reference works covering different aspects of power electronics, such as Variable Frequency Power Supply, DC Power Supply, Magnetic Technology, New Energy Power Conversion, Electromagnetic Compatibility as well as Wireless Power Transfer Technology and Equipment. The series features leading Chinese scholars and researchers and publishes authored books as well as edited compilations. It aims to provide critical reviews of important subjects in the field, publish new discoveries and significant progress that has been made in development of applications and the advancement of principles, theories and designs, and report cutting-edge research and relevant technologies. The CPSS Power Electronics series has an editorial board with members from the China Power Supply Society and a consulting editor from Springer.

Readership: Research scientists in universities, research institutions and the industry, graduate students, and senior undergraduates.

More information about this series at http://www.springer.com/series/15422

Wenxing Zhong · Dehong Xu ·
Ron Shu Yuen Hui

# Wireless Power Transfer

## Between Distance and Efficiency

Springer

Wenxing Zhong
College of Electrical Engineering
Zhejiang University
Hangzhou, Zhejiang, China

Dehong Xu
College of Electrical Engineering
Zhejiang University
Hangzhou, Zhejiang, China

Ron Shu Yuen Hui
The University of Hong Kong
Hong Kong, China

ISSN 2520-8853 ISSN 2520-8861 (electronic)
CPSS Power Electronics Series
ISBN 978-981-15-2440-0 ISBN 978-981-15-2441-7 (eBook)
https://doi.org/10.1007/978-981-15-2441-7



This Springer imprint is published by the registered company Springer Nature Singapore Pte Ltd.
The registered company address is: 152 Beach Road, #21-01/04 Gateway East, Singapore 189721, Singapore

# Preface

Human being never stops chasing freedom, from wheels to aircrafts, from corded phones to mobiles, from manual work to automation, and so on and so forth. One century ago, Tesla tried to demonstrate the freedom of power transmission and laid down the basic principles for both inductive and capacitive wireless power transfer (WPT). Unfortunately, there were few WPT applications at the time and the undeveloped power electronics also set limitations for implementing these brilliant ideas. Nowadays, the development of EVs, transportation electrification, self-driving techniques, unmanned moving devices, and portable electronics is forming great demand for free powering. Cords and manual operation get in the way of efficiency in these areas. Automatic and safe power transmission or charging technologies for these applications will be surely accomplished in the near future.

The nature of WPT is freedom and the power transfer distance is the key index. Radio has a long transfer distance and so it gives freedom to information transformation. Radio frequency (RF) is now under investigation for power transformation. Though it seems difficult to conquer the divergence challenge, feasible power transfer solutions have been developed based on RF for applications with low power consumptions. For higher power levels, magnetic resonance coupling is leading other approaches.

Another critical feature for medium and high power WPT is the energy efficiency. There is always a contradiction between the power transfer distance and the power transfer efficiency. A higher efficiency not only means less energy waste but also offer a larger freedom of operation. Higher efficiency, by the way, relieves engineers from the anxiety of dealing with the heat and turns bulky electrical circuits and components into elegant bricks.

This book introduces the background and basic theory of magnetic resonance WPT, and summarizes some research outcomes on multi-resonator wireless power transfer and maximum-efficiency-operation (MEO) techniques. We hope this book can serve as a technical guideline or reference for engineers and researchers working on inductive WPT.

Hangzhou, China Wenxing Zhong
Hangzhou, China Dehong Xu
Hong Kong, China Ron Shu Yuen Hui

# Contents

# Part I
# Fundamentals of Magnetic Resonance Wireless Power Transfer

# Chapter 1
# Introduction to Magnetic Resonance WPT

The scientific breakthroughs of the French scientist André-Marie Ampère (1775–1836) and English scientist Michael Faraday (1791–1867) in early nineteenth century laid down the foundation of electricity and electromagnetism. The Ampere's law, Faraday's law, and electromagnetism form the pillars of modern electrical engineering. Their work was further enhanced by Scottish scientist James Clerk Maxwell (1831–1879) who developed the well-known Maxwell's equations for electromagnetic fields and waves. These scientific principles became the tools for early wireless power transfer (WPT) pioneers, such as Tesla, Hutin, and Leblanc [1–3], to explore WPT through the use of coupled alternating magnetic field.

Among early WPT pioneers, Nicola Tesla was undoubtedly the most influential scientist in WPT research. Tesla was recognized as a highly visionary inventor who was well ahead of his time [4, 5]. His many inventions such as AC machines, AC power generation and transmission, radio transmitters and receivers, X-ray machines and the world's first remote-controlled technology have been used for over a century. Although some of his inventions have been improved over time with the availability of more advanced technologies, their basic concepts are still being used every day in modern societies.

Between the WPT research of Tesla in late nineteenth century and mid 1980s, while research in WPT was still active in medical implants, commercial applications of WPT was not wide spread. Tesla pointed out that WPT could take place through magnetic coupling at high frequency and under resonance conditions. To maintain high energy efficiency, WPT requires (i) windings with low resistance at high-frequency operation and (ii) a high-frequency power source. While Litz wires have become available since 1950s, high-frequency switching power electronics switches (namely power MOSFETs) only became commercially available in 1980s. Such historical developments of Litz wires and power electronics explain why research in magnetic resonance-based WPT took off for medical implants [6, 7] in late 1980s and electric vehicles and inductive power pickups in mid 1990s [8, 9]. The dawn of the mobile phone era in mid 1990s also prompted active research in wireless charging for portable electronics [10–15].


W. Zhong et al., *Wireless Power Transfer*, CPSS Power Electronics Series,
https://doi.org/10.1007/978-981-15-2441-7_1

Apart from magnetic resonance coupling, many alternative WPT techniques have been explored, such as microwave power transmission [16], laser power transmission [17], capacitive power transmission [18], etc. However, it seems that for the applications with medium to high power levels, the approach based on magnetic resonance coupling invented by Tesla outperforms the other techniques in terms of power transfer capability, power transfer efficiency, safety, easy-to-control, etc. Therefore, most of the commercialized wireless charging platforms for mobile phones and other portable electronics adopt the magnetic resonance WPT. Another application that might adopt magnetic resonance WPT is the wireless charging of electric vehicles (EVs). As many countries announced their plans to ban internal-combustion engine vehicles, a transition to electric vehicles (EVs) seems to be inevitable. While another sure trend is the commercialization of the self-driving vehicle technology. It is natural that self-driving EVs are charged automatically. Therefore wireless charging is a suitable choice. A brief review on magnetic resonance WPT is provided in this chapter, including several important milestones. More history of WPT could be found in [19–24].

## 1.1 Nikola Tesla's Early Work

Since the end of the nineteenth century, "resonance" was a common theme in several aspects of Tesla's work. In 1891, he patented the well-known "Tesla coil" in which the concept of *oscillation coupling* was implemented, shown in Fig. 1.1. The high voltage transformer will firstly charge up the capacitor until its voltage reaches the breakdown voltage of the spark gap. Then the spark gap will form a low-resistance conducting path and thus, the primary inductor and the capacitor will oscillate at the resonant frequency typically in the range of 50 kHz–1 MHz [25]. Although there is no lump capacitor in the receiver side, there exists parasitic capacitance in the secondary winding and also capacitance between the torus and the ground. These capacitances will form another oscillation in the receiver side. This is exactly the basic operation principle of the magnetic resonance WPT, which will be explained in detail in the

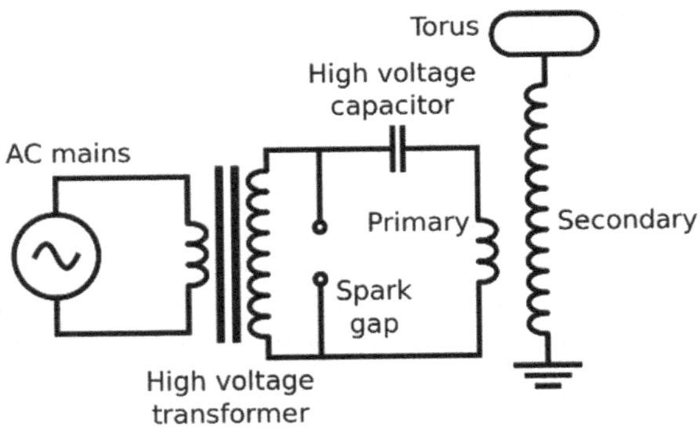


**Fig. 1.1** Schematic of the typical structure of a Tesla coil [25] (*Source file* https://commons.wikimedia.org/wiki/File:Tesla_coil_3.svg)

next chapter. From the research of Tesla to the early 1990s, WPT was barely adopted in commercial applications. But the research in WPT was still active, such as, John Schuder and his team reported the use of a pair of magnetically coupled circuits for WPT through the chest wall of an animal in 1961 [26] and they succeeded in transferring 1 kW through the skin of a dog in 1971 [27], Fuller proposed the bipolar transistor-based astable multivibrator for WPT through a tissue barrier in 1968 [28], Bo Cho used a pair of magnetically coupled circuits operating at a resonance mode for a transcutaneous energy transmission system in 1990 [29]. More details could be found in [20].

## 1.2 Inductive Power Transfer (IPT)

Although the concept of magnetic resonance coupling was established by Tesla before 1890s, it has not been successfully applied to industrial applications in the first half of the twentieth century. Wireless power transfer based on magnetic resonance regained attention of researchers for biomedical applications in the 1960s. The maturity of power electronics in 1980s offered a solution to develop high frequency power supplies which were essential to WPT. In the 1980s, Boys and Green used power inverters and magnetic resonance for IPT applications [30–32], while Hui developed planar WPT technologies for portable electronics. The schematic of the IPT system is shown in Fig. 1.2. This system consists of two mutually coupled windings, i.e., the track (a large conducting loop consisting of two parallel paths with opposite current directions) and the pickup winding. The pickup is able to move with respect to the track. Despite the fact that the primary winding is a long track, the system realizes WPT through the same mechanism of magnetic resonance coupling. Without the constraint of power cords, it was found that the IPT system was a good solution to factory automation and clean factories [21, 32].

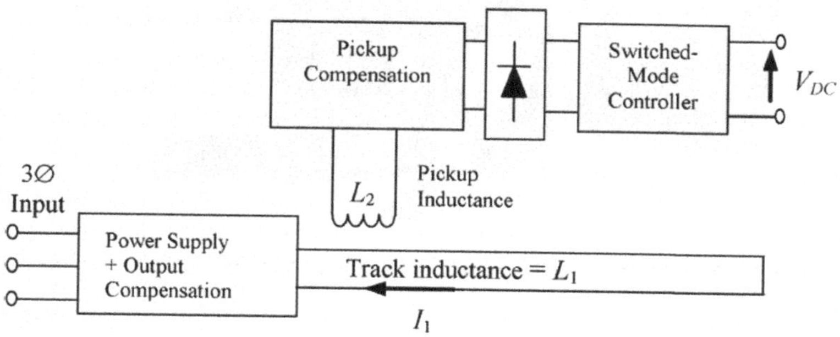


**Fig. 1.2** Schematic of the IPT system [32]

## 1.3 Planar Wireless Charging Technology and Qi

From the 2000s, magnetic resonance coupling found applications in the wireless charging of portable electronics. It was reported that a huge amount of electronic wastes was produced by the various adaptors of mobile phones and other portable electronic products since 1980s [33, 34]. The reason is simply that each product and each generation of the product will have its own charger. Therefore, adopting a common charging protocol for all kinds of portable electronic products is a sound solution to this issue. In 2008, the Wireless Power Consortium (WPC) was formed and in 2010 the first version of the wireless charging standard Qi was released. Figure 1.3 shows a Qi wireless charging pad with a variety of compatible and noncompatible items. By the end of 2018, Qi standard has been supported by more than 600 companies and incorporated into more than 3000 products [35].

Adopting the principle of magnetic resonance coupling, Qi standard is based on a planar wireless charging technology with the "free-positioning" and "localized charging" features [36]. Figure 1.4 shows the top-view of a Qi charging pad where the black dots indicate the positions of the ferrite rods for the primary coils, the ring surrounding each ferrite rod is the transmitting winding and the large circles represent the profiles of the receiving coils. Free positioning means that the receiving coil can be placed anywhere on the pad and it will always cover at least one transmitting coil and thereby, localized charging which means only the transmitting coil (or coils) underneath the receiving coil is/are energized, is enabled [37].

**Fig. 1.3** A planar wireless charging pad for portable electronics [38]

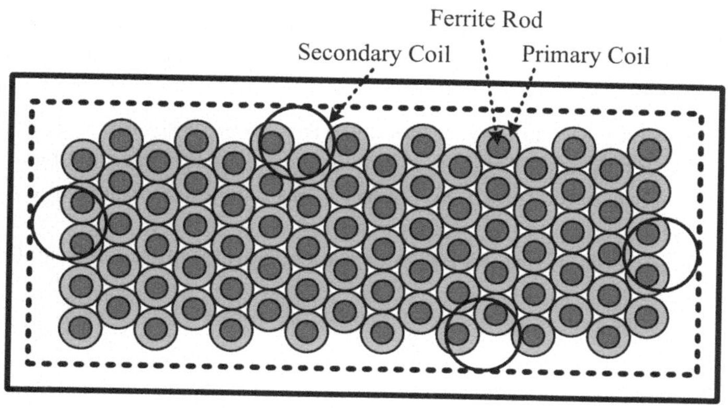


**Fig. 1.4** A Qi charging pad with the "free-positioning" and "localized charging" features

## 1.4 The Four-Coil System

In 2007, the four-coil structure was proposed in [39]. This structure involves a Power driving coil, a Sending resonator, a Receiving resonator, and a Load coil as shown in Fig. 1.5. The use of the power driving coil and the load coil offer two extra mutual coupling coefficients for impedance matching (assuming that the mutual coupling of the driving loop and the load loop is negligible). Besides the mutual coupling coefficient between the Sending resonator and the Receiving resonator ($k_{SR}$), the two extra coefficients are the coupling coefficient between the Power driving coil and the Sending resonator ($k_{PS}$), and the coupling coefficient between the Receiving resonator and the Load coil ($k_{RD}$). Reference [40] provides a detailed circuit analysis of this four-coil system and step-by-step explanations on how to match impedances in various stages in order to maximize power transfer. The four-coil system provides three mutual coupling coefficients which can be utilized to maximize the power transfer range at the expense of system energy efficiency [41].

**Fig. 1.5** Schematic of the four-coil WPT system

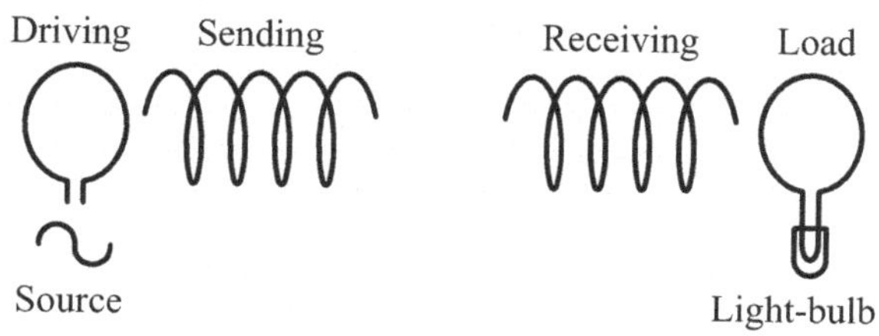

## 1.5 Summary

Despite the original concept was developed over a century ago, WPT is still a fast growing R&D area with immense application potentials. With the understanding of the historical WPT developments, it is hoped that visionary researchers would take WPT to new heights in the future. So far, WPT has reached commercialization stage in consumer electronics and manufacturing industry. It is envisaged that the next sectors that would benefit from WPT include electric vehicles and medical implants.

## References


1. Tesla N (1898) High frequency oscillators for electro-therapeutic and other purposes. Electr Eng XXVI(550)
2. Tesla N (1904) Transmission of electrical energy without wire. Electrical World and Engineer, 5 March 1904 [online]. www.tfcbooks.com/tesla/
3. Hutin M, Leblanc M (1894) Transformer system for electric railways. US patent 527 857, 23 Oct 1894
4. "Nikola Tesla 1857–1943", May 1943. In: Proceedings of the I.R.E., 194 pp
5. Lomas R (1999) The man who invented the twentieth century—Nikola Tesla—forgotten genius of electricity. Headline Book Publishing, UK
6. Joung GB, Cho BH (1998) An energy transmission system for an artificial heart using leakage inductance compensation of transcutaneous transformer. IEEE Trans Power Electron 13(6):1013–1022
7. Ghahary A, Cho BH (1990) Design of a transcutaneous energy transmission systems using a series resonant converter. In: IEEE power electronics specialists conference, pp 1–8
8. Green AW, Boys JT (1994) 10 kHz inductively coupled power transfer—concept and control. In: Power electronics and variable speed drives conference, pp 694–699
9. Elliott GAJ, Boys JT, Green AW (1995) Magnetically coupled systems for power transfer to electric vehicles. In: Proceedings of 1995 international conference on power electronics and drive systems, vol 2, pp 797–801
10. Jang Y, Jovanovic M (2003) A contactless electrical energy transmission system for portable-telephone battery chargers. IEEE Trans Ind Electron 50(3):520–527
11. Kim C-G, Seo D-H, You J-S, Park J-H, Cho BH (2001) Design of a contactless battery charger for cellular phone. IEEE Trans Ind Electron 48(6):1238–1247
12. Jang Y, Jovanovic MM (2000) A contactless electrical energy transmission system for portable-telephone battery chargers. In: Twenty-second international telecommunications energy conference, INTELEC, pp 726–732
13. Choi B, Nho J, Cha H, Ahn T, Choi S (2004) Design and implementation of low-profile contactless battery charger using planar printed circuit board windings as energy transfer device. IEEE Trans Ind Electron 51(1):140–147
14. Hui SYR (2005) Planar inductive battery charger. UK Patent GB2389720B, 7 Sept 2005
15. Cheng L, Hay JW, Beart P (2005) Contact-less power transfer. US Patent 6,906,495, 14 June 2005
16. Brown WC (1964) The history of power transmission by radio waves. IEEE Trans Microw Theory Tech MTT-32(9):1230–1242
17. Jin K, Zhou W. Wireless laser power transmission: a review of recent progress. IEEE Trans Power Electron
18. Hu AP, Liu C, Li HL (2008) A novel contactless battery charging system for soccer playing robot. In: 2008 15th international conference on mechatronics and machine vision in practice, Auckland, pp 646–650

19. Sun T, Xie X, Wang Z (2013) Wireless power transfer for medical microsystems. Springer, New York
20. Hui SYR (2016) Magnetic resonance for wireless power transfer [A look back]. IEEE Power Electron Mag 3(1):14–31
21. Boys JT, Covic GA (2015) The inductive power transfer story at the University of Auckland. IEEE Circuits Syst Mag 15(2):6–27 (second quarter 2015)
22. Dai J, Ludois DC (2015) A survey of wireless power transfer and a critical comparison of inductive and capacitive coupling for small gap applications. IEEE Trans Power Electron 30(11):6017–6029
23. Brown WC (1996) The history of wireless power transmission. Sol Energy 56(1):3–21
24. Garnica J, Chinga RA, Lin J (2013) Wireless power transmission: from far field to near field. Proc IEEE 101(6):1321–1331
25. On line: https://en.wikipedia.org/wiki/Tesla_coil
26. Schuder JC, Stephenson HE, Townsend JF (1961) High level electromagnetic energy transfer through a closed chest wall. IRE Int Conv Rec 9:119–126
27. Schuder JC, Gold JH, Stephenson HE (1971) An inductively coupled RF system for the transmission of 1 kW of power through the skin. IEEE Trans Biomed Eng 18(4):265–273
28. Apparatus for efficient power transfer through a tissue barrier. IEEE Trans Biomed Eng 15(1):63–65
29. Ghahary A, Cho B (1990) Design of a transcutaneous energy transmission system using a series resonant converter. In: Proceedings of 21st annual IEEE Power Electronics Specialists Conference, pp. 1–8
30. Boys JT, Green AW (1994) Inductive power distribution system. U.S. Patent 5293308A, 8 Mar 1994
31. Wang C-S, Covic GA, Stielau OH (2004) Power transfer capability and bifurcation phenomena of loosely coupled inductive power transfer systems. IEEE Trans Ind Electron 51(1):148–157
32. Covic GA, Boys JT (2013) Inductive power transfer. Proc IEEE 101(6):1276–1289
33. Leung A, Luksemburg W, Wong A, Wong M (2007) Spatial distribution of polybrominated diphenyl ethers and polychlorinated dbenzo-p-dioxins and dibenzofurans in soil and combusted residue at Guiyu, an electronic waste recycling site in southeast China. Environ Sci Technol 41(8):2730–2737
34. Wong CSC, Duzgoren-Aydin NS, Aydin A, Wong MH (2007) Evidence of excessive releases of metals from primitive e-waste processing in Guiyu, China. Environ Pollut 17240013E
35. Wireless Power Consortium (2012) [Online]. http://www.wirelesspowerconsortium.com
36. Liu X, Ho WC, Hui SYR, Chan WC (2011) Localized charging, load identification and bi-directional communication methods for a planar inductive battery charging system. U.S. Patent 7 915 858, 29 Mar 2011
37. Zhong WX, Liu X, Ron Hui SY (2011) A novel single-layer winding array and receiver coil structure for contactless battery charging systems with free-positioning and localized charging features. IEEE Trans Ind Electron 58(9):4136–4144
38. Hui SY (2013) Planar wireless charging technology for portable electronic products and Qi. Proc IEEE 101(6):1290–1301
39. Kurs A, Karalis A, Moffatt R, Joannopoulos JD, Fisher P, Soljacˇic´ M (2007) Wireless power transfer via strongly coupled magnetic resonances. Science 317(5834):83–86
40. Cheon S, Kim YH, Kang SY, Lee ML, Lee JM, Zyung T (2011) Circuit-model-based analysis of a wireless energy-transfer system via coupled magnetic resonances. IEEE Trans Ind Electron 58(7):2906–2914
41. Hui SYR, Zhong W, Lee CK (2014) A critical review of recent progress in mid-range wireless power transfer. IEEE Trans Power Electron 29(9):4500–4511

# Chapter 2
# Basic Theory of Magnetic Resonance WPT

In this book, small italic letters are used to represent time-domain variables, bold letters to represent phasors and capital italic letters to represent real numbers or root mean square (RMS) values. For example, $i_1$ is an instantaneous current variable which changes with time; $\mathbf{I_1}$ is a current phasor and $I_1$ is the RMS value of $\mathbf{I_1}$.

It should be noted that the terminology of "efficiency" in radio-frequency (RF) field and power electronics field may be different [1]. For RF researchers, to maximize the efficiency may refer to maximizing the output power for a given input voltage. For power electronics researchers, to maximize efficiency is equivalent to minimizing the energy losses for a given output power, or in other words *efficiency* is defined as the total output power divided by the total input power of a system. In this book, the discussion is focused on the latter definition commonly used by the power electronics community.

The fundamental operation principles of magnetic resonance WPT will be given in this chapter, which starts with the introduction to WPT of two coupled inductors. Then it is proved that both the power transfer capability and the efficiency can be improved by properly compensating the coupled inductors. Later, the characteristics of a typical series-compensated primary and secondary (SS) WPT system are discussed in detail.

## 2.1 From Coupled Inductors to Magnetic Resonance Coupling

The core part of an inductive WPT system is the magnetically coupled windings, which can be modeled with two coupled inductors, as shown in Fig. 2.1. The corresponding circuit equations are given by

$$(R_1 + j\omega L_1)\mathbf{I_1} + j\omega M_{12}\mathbf{I_2} = \mathbf{V_S} \tag{2.1}$$

$$j\omega M_{12}\mathbf{I_1} + (R_2 + R_L + j\omega L_2)\mathbf{I_2} = \mathbf{0}, \tag{2.2}$$


W. Zhong et al., *Wireless Power Transfer*, CPSS Power Electronics Series,
https://doi.org/10.1007/978-981-15-2441-7_2

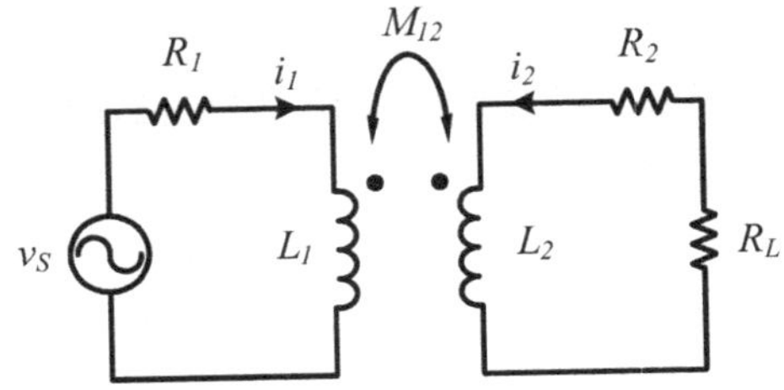


**Fig. 2.1** The coupled inductors

where $R_1$ and $R_2$ are the resistance of the primary and secondary windings, respectively; $L_1$ and $L_2$ are the self-inductances of the primary and secondary windings, respectively; $M_{12}$ is the mutual inductance between two windings; $R_L$ is the equivalent load resistance.

The secondary current phasor $\mathbf{I_2}$ can be derived from (2.2) as

$$\mathbf{I_2} = \frac{-j\omega M_{12}\mathbf{I_1}}{R_2 + R_L + j\omega L_2}. \tag{2.3}$$

For a given primary current $I_1$, it produces a loss of $I_1^2 R_1$ in the primary winding and induces a voltage with the RMS value of $\omega M_{12} I_1$ in the secondary winding. This induced voltage will apply on the impedance $R_2 + R_L + j\omega L_2$ in the secondary side. So the induced current will be largely restricted by the reactance of the secondary winding. If the reactance of the secondary side impedance could be compensated, for example, by connecting a capacitor $C_2$ in series with the secondary winding and canceling out the reactance as shown in Fig. 2.2, i.e.,

$$\omega L_2 - \frac{1}{\omega C_2} = 0 \tag{2.4}$$

then a larger load current and a higher load power can be transferred to the load. With the same input current in the primary winding, the output power is increased and thereby, the power transfer capability and efficiency are improved. In the primary side, $j\omega L_1$ should be also compensated by using capacitors to minimize the required VA rating of the system, i.e., a lower input voltage is required to generate the same

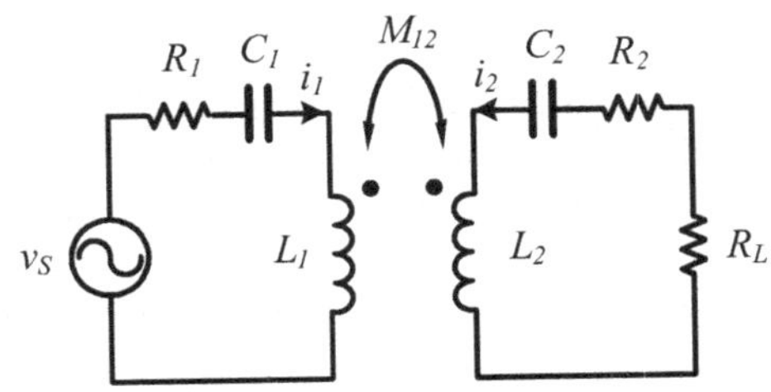


**Fig. 2.2** Coupled inductors with compensating capacitors connected in series with the windings

input primary winding current. Therefore, primary compensation also improves the power transfer capability of a WPT system.

One of the basic operation principles of magnetic resonance WPT is to compensate both the primary and secondary reactances and thus, we have resonance in both the primary and secondary sides. A series-compensated primary and secondary system is shown in Fig. 2.2. There are two L-C (inductor and capacitor) resonators in the circuit. This forms a basic magnetic resonance WPT system.

Besides using series compensation in both the primary and secondary sides which is usually called series–series (SS) compensation, there is also parallel compensation (denoted with letter P). Consequently, there are totally four basic compensation topologies, i.e., SS, SP, PS, and PP [1], as shown in Fig. 2.3. For example, PP means both primary and secondary sides use parallel compensation. Regardless of the different connections, series and parallel compensations share the same purpose of eliminating the reactances in both the primary and secondary circuits. So the following analyses will focus on SS topology. Similar studies can be extended to the other three topologies.

The circuit model of the SS WPT system is given by

$$(R_1 + jX_1)\mathbf{I_1} + j\omega M_{12}\mathbf{I_2} = \mathbf{V_S} \tag{2.5}$$

$$j\omega M_{12}\mathbf{I_1} + (R_2 + R_L + jX_2)\mathbf{I_2} = \mathbf{0}, \tag{2.6}$$

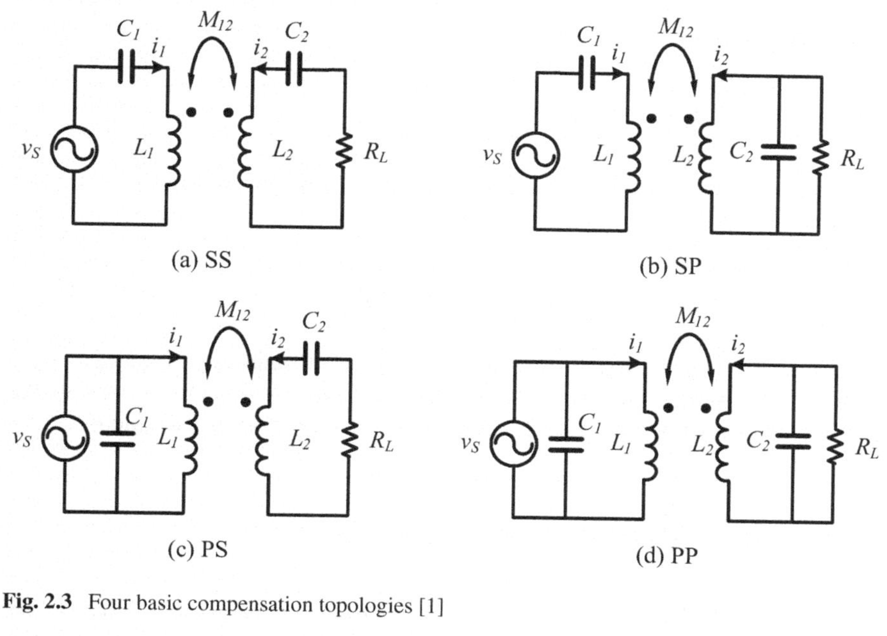


**Fig. 2.3** Four basic compensation topologies [1]

where $X_i = \omega L_i - 1/(\omega C_i)$, $i = 1$ or 2, is the reactance of Resonator-$i$, i.e., Here $R_i$ include both the winding resistance and the equivalent series resistance (ESR) of the compensating capacitor.

## 2.2 Characteristics of an SS WPT System

### 2.2.1 *Reflected Impedance [2]*

From (2.6), we get

$$\mathbf{I_2} = \frac{-j\omega M_{12}\mathbf{I_1}}{R_2 + R_L + jX_2}. \tag{2.7}$$

By substituting (2.7) into (2.5), we get the input impedance of the system as

$$\mathbf{Z_{in}} = \frac{\mathbf{V_S}}{\mathbf{I_1}} = R_1 + jX_1 + \frac{\omega^2 M_{12}^2}{R_2 + R_L + jX_2}. \tag{2.8}$$

The first term, $(R_1 + jX_1)$, is the impedance of the primary circuit. The second term is the reflected impedance, namely, the additional equivalent impedance seen by the power source due to the current in the receiving winding [2]. So the reflected impedance is given by

$$Z_r = \frac{\omega^2 M_{12}^2}{R_2 + R_L + jX_2}. \tag{2.9}$$

At the resonant frequency of the secondary resonator at which $X_2 = 0$, (2.9) becomes

$$Z_r = R_r = \frac{\omega^2 M_{12}^2}{R_2 + R_L}. \tag{2.10}$$

Therefore, when the system operates at the resonant frequency of the secondary resonator, the reflected impedance has only the resistive part and it will not bring any reactance in the primary circuit. This is a unique feature of SS WPT systems.

### 2.2.2 *Induced Voltages, Currents and Gains*

It is assumed that the resonant frequencies of both the primary and secondary resonators are the same and they are equal to the operating frequency of the system.

Therefore, $X_1$ and $X_2$ in (2.5) and (2.6) are zero. The induced voltage and current in the secondary side can be derived, respectively, as

$$\mathbf{V}_{2\_\text{induced}} = j\omega M_{12}\mathbf{I}_1 \tag{2.11}$$

$$\mathbf{I}_2 = \frac{-j\omega M_{12}\mathbf{I}_1}{R_2 + R_L}. \tag{2.12}$$

Therefore, the secondary current phasor $\mathbf{I_2}$ is 90° lagging behind the primary current phasor $\mathbf{I_1}$. From (2.12), the current gain is derived as

$$\frac{I_2}{I_1} = \frac{\omega M_{12}}{R_2 + R_L}. \tag{2.13}$$

If the parasitic resistance in the secondary side is neglected, (2.13) becomes

$$\frac{I_2}{I_1} = \frac{\omega M_{12}}{R_L}. \tag{2.14}$$

Also, if the parasitic resistance is neglected, the load voltage equals to the induced voltage, i.e.,

$$V_L = \omega M_{12} I_1. \tag{2.15}$$

Therefore, if the primary winding current is maintained constant, the series-compensated secondary is a voltage source and the output voltage is independent of the load. In the primary side, the induced voltage is

$$\mathbf{V}_{1\_\text{induced}} = j\omega M_{12}\mathbf{I}_2. \tag{2.16}$$

If the parasitic resistance in the primary side is neglected, the source voltage equals to the induced voltage, i.e.,

$$V_S = \omega M_{12} I_2. \tag{2.17}$$

From (2.15), (2.17), and (2.14), the voltage gain can be obtained as

$$\frac{V_L}{V_S} = \frac{I_1}{I_2} = \frac{R_L}{\omega M_{12}}. \tag{2.18}$$

This simplified equation is especially useful when designing a WPT system with given input and output voltages. With a given operating frequency, equivalent load resistance and voltage ratio, the required mutual inductance can be obtained.

### 2.2.3 Efficiency

Because inductive WPT is based on the near-field coupling, the radiation loss is negligible and all the losses of the coupled resonators will be on the parasitic resistances of the resonators, i.e., $R_1$ and $R_2$. Therefore, the efficiency of the SS system in Fig. 2.2 can be expressed as

$$\eta = \frac{\text{total output power}}{\text{total input power}} = \frac{I_2^2 R_L}{I_1^2 R_1 + I_2^2 (R_2 + R_L)} = \frac{R_L}{\left(\frac{I_1}{I_2}\right)^2 R_1 + R_2 + R_L}. \quad (2.19)$$

The ratio $I_1/I_2$ is determined solely by (2.6), which implies that the efficiency of the system will not be affected by (2.5). Thereby, the compensation condition in Resonator-1, i.e., the value of $X_1$, will not affect the efficiency of the system. With (2.6), the efficiency can be worked out as

$$\eta = \frac{\omega^2 M_{12}^2 R_L}{[(R_2 + R_L)^2 + X_2^2] R_1 + \omega^2 M_{12}^2 (R_2 + R_L)}. \quad (2.20)$$

Equation (2.20) implies that for a system with given coils, the larger the $M_{12}$, the higher the efficiency. By solving the following equation,

$$\frac{\partial \eta}{\partial X_2} = 0 \quad (2.21)$$

the optimal value of $X_2$ for maximizing the efficiency is obtained as

$$X_{2\_\mathrm{OPT}} = 0. \quad (2.22)$$

This implies that for a given operating frequency, the optimal series compensating capacitance in the secondary circuit is the one that forms an LC resonance at this operating frequency. In this calculation, the ESR of the capacitor can be assumed unchanged because when the capacitance changes, the effect on the efficiency is mainly due to the reactance change rather than the ESR change.

By substituting (2.22) into (2.20) and solving the following equation

$$\frac{\partial \eta}{\partial R_L} = 0 \quad (2.23)$$

the optimal values $R_L$ can be obtained as

$$R_{L\_\mathrm{OPT}\_\eta} = R_2 \sqrt{1 + \frac{\omega^2 M_{12}^2}{R_1 R_2}} = R_2 \sqrt{1 + k_{12}^2 Q_1 Q_2}, \quad (2.24)$$

where $k_{12} = \frac{M_{12}}{\sqrt{L_1 L_2}}$ is the coupling coefficient; $Q_i = \frac{\omega_0 L_i}{R_i}$ ($i = 1$ or 2) is the quality factor or unloaded quality factor of Resonator-$i$ and $\omega_0$ is the resonant frequency of the secondary resonator. The maximum efficiency of the system can then be given by

$$\eta_{\max} = \frac{k_{12}^2 Q_1 Q_2}{\left(1 + \sqrt{1 + k_{12}^2 Q_1 Q_2}\right)^2}. \quad (2.25)$$

In some applications, $k_{12}^2 Q_1 Q_2$ will be much larger than 1. For example in EV charging applications, $k_{12}$ is normally larger than 0.1, and $Q_1$ and $Q_2$ are typically larger than 100, and thus $k_{12}^2 Q_1 Q_2$ is larger than 100. Under this assumption,

$$R_{L\_OPT\_\eta} = R_2\sqrt{1 + k_{12}^2 Q_1 Q_2} \approx R_2 k_{12}\sqrt{Q_1 Q_2}. \quad (2.26)$$

Moreover, if $Q_1$ is equal to or close to $Q_2$ as in many cases identical windings are used in both the primary and secondary sides, (2.26) becomes

$$R_{L\_OPT\_\eta} \approx R_2 k_{12} Q_2 = k_{12}\omega_0 L_2. \quad (2.27)$$

From (2.25), one can see that the larger the $k_{12}^2 Q_1 Q_2$, the higher the maximum efficiency. Therefore, the maximum achievable efficiency of a two-resonator SS WPT system depends on $k_{12}^2 Q_1 Q_2$ which is also known as the "$kQ$ product" [3]. Normally, $k$ is more or less fixed in a specific application with a given power transfer distance and given dimensions for the windings. While windings can be optimized for higher efficiency and better tolerance for load variations. Figure 2.4 shows the efficiency of an SS WPT system under load resistance and quality factor changes. Here, the operating frequency is 100 kHz, the inductance of the windings is 100 μH, and the coupling coefficient is 0.2. The load resistance is normalized to the optimal load

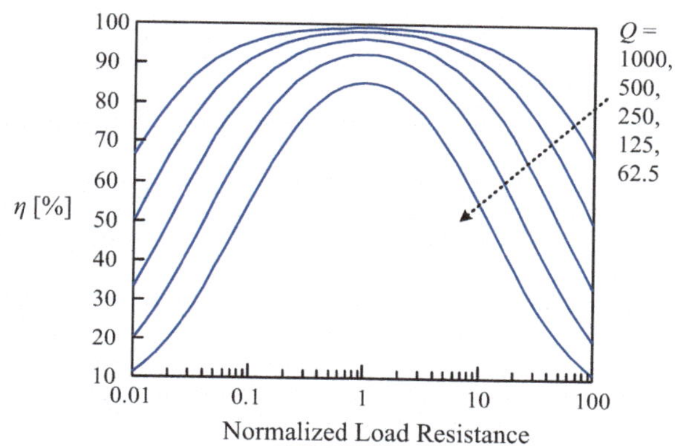


**Fig. 2.4** Efficiency of an SS WPT system as a function of the load resistance and the quality factor of the windings

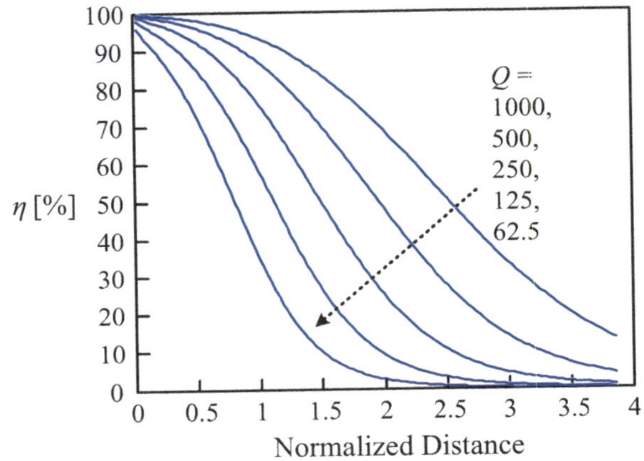


**Fig. 2.5** Maximum efficiency of an SS WPT system as a function of the power transfer distance and the quality factor of the windings

resistance given by (2.24). Larger quality factor improves the efficiency as expected. Besides, it should be noted that the efficiency improvement is more significant at nonoptimal load resistance points than at the optimal load resistance point. Figure 2.5 shows the typical maximum efficiency curves of an SS WPT system as a function of the power transfer distance and $Q$ when identical windings are used and the load resistance is optimized for each case. The power transfer distance is normalized to the diameter of the circular windings. With a larger quality factor, the maximum efficiency decreases slower as the transfer distance increases. The parameters used in this plot are given in Table 4.1.

On the other hand, the efficiency of a WPT system can be expressed as the product of the efficiencies of the primary and secondary sides [4], i.e.,

$$\eta = \eta_P \eta_S, \tag{2.28}$$

where $\eta_P$ is the efficiency of the primary side which is given by

$$\eta_P = \frac{\mathrm{Re}(Z_r)}{R_1 + \mathrm{Re}(Z_r)} \tag{2.29}$$

and $\eta_S$ is the efficiency of the secondary side which is given by

$$\eta_S = \frac{R_L}{R_2 + R_L}. \tag{2.30}$$

Note that at the secondary resonant frequency, $\mathrm{Re}(Z_r)$ is given by

$$\mathrm{Re}(Z_r) = \frac{\omega_0^2 M^2}{R_L}. \tag{2.31}$$

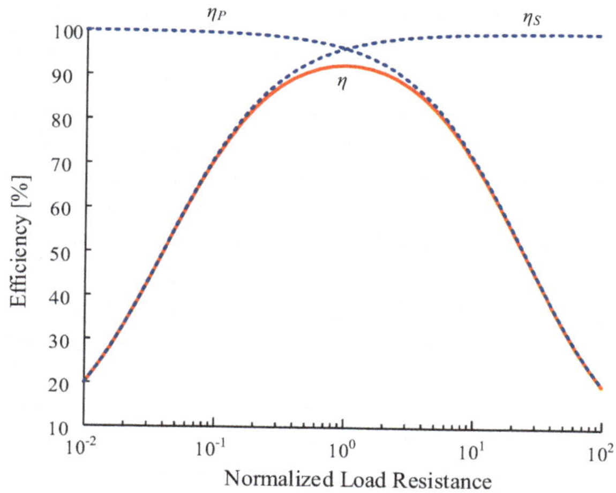


**Fig. 2.6** Efficiency curves of the primary side, the secondary side, and the whole system of an SS WPT system

Therefore, when the load resistance increases, the secondary efficiency will increase while the reflected resistance in the primary side will decrease according to (2.31) which lowers primary efficiency. Figure 2.6 shows the efficiency curves of the primary side, secondary side and whole system of the same SS WPT system as in Fig. 2.4 and the quality factor is set at 125. The load resistance is normalized to the optimal load resistance given by (2.24). It is observed that the maximum efficiency can be achieved only when the losses of two sides are properly balanced.

### *2.2.4 Output Power*

From the basic model of an SS WPT system and assuming the system are perfectly tuned, i.e., the primary and secondary resonators are resonant at the operating frequency, the output power can be obtained as

$$P_O = \frac{V_S^2 \omega^2 M_{12}^2 R_L}{(\omega^2 M_{12}^2 + R_1 R_2 + R_1 R_L)^2}. \tag{2.32}$$

The optimal load resistance for maximizing the output power can be derived as

$$R_{L\_\mathrm{OPT}\_P_O} = R_2\left(1 + \frac{\omega^2 M_{12}^2}{R_1 R_2}\right). \tag{2.33}$$

Comparing (2.33) and (2.24) leads to the following inequality

$$R_{L\_\mathrm{OPT}\_\eta} < R_{L\_\mathrm{OPT}\_P_O} \tag{2.34}$$

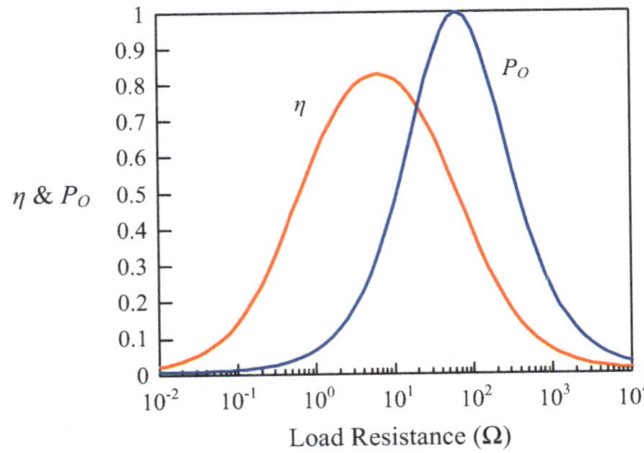


**Fig. 2.7** Efficiency and output power curves of an SS WPT system with a constant source voltage

Figure 2.7 shows the efficiency and output power curves versus the load resistance of an SS WPT system where a constant source voltage is assumed.

### 2.2.5 *Input Impedance, Zero Phase Angle, and Bifurcation [5]*

Recall that the input impedance of an SS system is the sum of the primary impedance and the reflected impedance.

$$\mathbf{Z_{in}} = R_1 + jX_1 + Z_r. \tag{2.35}$$

In a practical implementation, the operating frequency is adjusted so that zero phase angle (ZPA) of the input impedance is achieved to minimize the VA rating and maximize the power transfer capability of the system. However, sometimes there are multiple ZPA frequencies. This is the bifurcation phenomenon [5]. It has been proved in [5] that the condition to ensure a unique ZPA frequency for an SS system is

$$Q_P > \frac{4Q_S^3}{4Q_S^2 - 1}, \tag{2.36}$$

where $Q_P$ and $Q_S$ are the loaded quality factors which are defined, with the parasitic resistances neglected, as

$$Q_P = \frac{\omega_0 L_1}{\omega_0^2 M^2 / R_L} \tag{2.37}$$

$$Q_S = \frac{\omega_0 L_2}{R_L} \tag{2.38}$$

and $\omega_0$ is the resonant frequency of the resonators, i.e.,

$$\omega_0 = \frac{1}{\sqrt{L_1 C_1}} = \frac{1}{\sqrt{L_2 C_2}}. \tag{2.39}$$

Alternatively, if the parameters of the resonators are considered fixed, the condition of the load resistance to ensure a unique ZPA frequency can be derived from (2.36) as

$$R_L > \omega_0 L_2 \sqrt{2 - 2\sqrt{1 - k_{12}^2}}. \tag{2.40}$$

Note that $k_{12}$ is normally much smaller than 1 and thereby

$$1 - k_{12}^2 \approx 1 - k_{12}^2 + \left(\frac{k_{12}^2}{2}\right)^2 = \left(1 - \frac{k_{12}^2}{2}\right)^2 \tag{2.41}$$

therefore, (2.40) can be rewritten as

$$R_L > k_{12}\omega_0 L_2. \tag{2.42}$$

From (2.27), $R_{L\_OPT\_\eta} \approx k_{12}\omega_0 L_2$ when the $kQ$ product is much larger than 1, and $Q_1$ and $Q_2$ are similar, so under this situation (2.42) can be rewritten as

$$R_L > R_{L\_OPT\_\eta} \tag{2.43}$$

Figure 2.8 shows the imaginary part of the input impedance of an SS WPT system as a function of the operating frequency and the load resistance. $k_{12}$ is 0.2 and

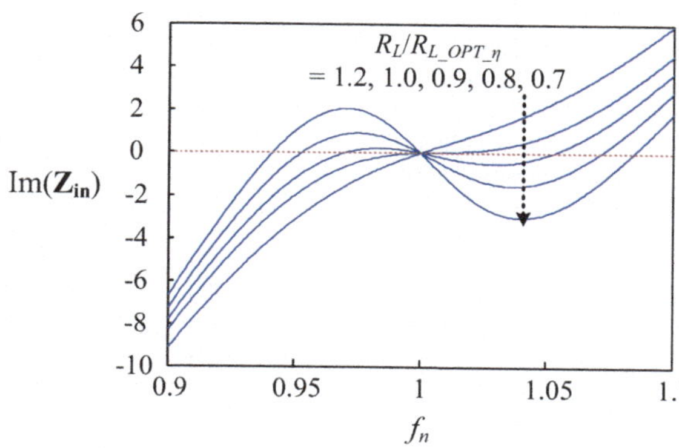


**Fig. 2.8** The imaginary part of the input impedance of an SS WPT system as a function of the operating frequency and the load resistance

$Q_1 = Q_2 = 250$. Note that the quality factors now are referred to the unloaded quality factors as previously defined. The operating frequency is normalized to the resonant frequency of the secondary resonator. When the load resistance is equal to or larger than the optimal value for maximum efficiency, ZPA is achieved only at the secondary resonant frequency. But when the load resistance is lower than the optimal value, there will be multiple ZPA frequencies. For a system with quality factors of 125, Fig. 2.9 shows the boundary load resistance (denoted as $R_{L\_b}$) for a unique ZPA point as defined by (2.40) and the optimal load resistance for maximum efficiency, varying with the coupling coefficient. One observation is that when $k_{12}$ is larger than some value (about 0.5 in this case study), the boundary load resistance will be larger than the optimal load resistance. This implies that when $k_{12}$ is larger than some value, bifurcation appears if the system operates at the maximum efficiency point. For the same secondary resonator with $Q_2$ of 125, Fig. 2.10 shows the curves of the optimal load resistance with different $Q_1$. While according to (2.40), the boundary load resistance is independent of $Q_1$. Therefore, if the quality factor of the primary resonator is smaller than that of the secondary resonator, the optimal load resistance

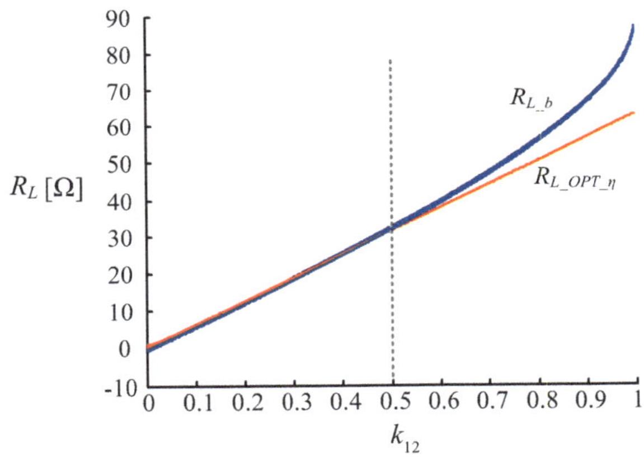


**Fig. 2.9** The boundary load resistance and the optimal load resistance of an SS WPT system as a function of the coupling coefficient

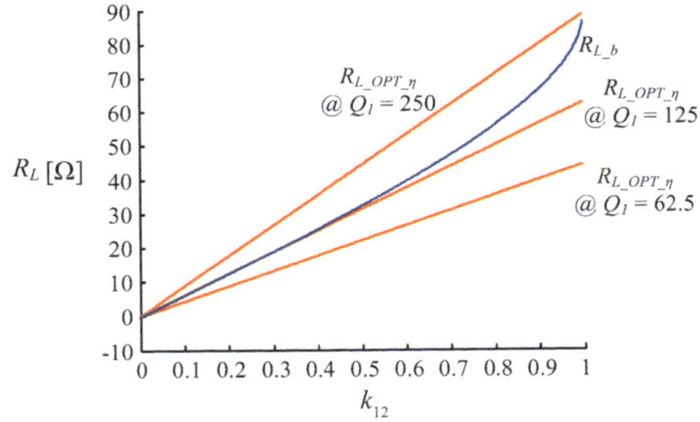


**Fig. 2.10** The boundary load resistance and the optimal load resistance of an SS WPT system as a function of the coupling coefficient and with different $Q_1$

will be lower than the boundary load resistance in the whole coupling coefficient range and thereby, bifurcation appears when operating at the maximum efficiency points. When the primary resonator has a larger quality factor, unique ZPA point can be achieved in a larger coupling range when operating at the maximum efficiency points.

## References

1. Stielau OH, Covic GA (2000) Design of loosely coupled inductive power transfer systems. In: Proceedings of the 2000 international conference on power system technology, vol 1, pp 85–90
2. Alexander CK, Sadiku MNO (2009) Fundamentals of electric circuits, 4th edn. McGraw-Hill, New York
3. Bosshard R, Kolar JW (2016) Multi-objective optimization of 50 kW/85 kHz IPT system for public transport. IEEE J Emerg Sel Top Power Electron 4(4):1370–1382
4. Li Y, Song K, Zhu C, Wei G, Lu R (2016) Efficiency optimizing and load matching analysis for the weak-coupling wireless power transfer system using a repeating coil. In: 2016 IEEE PELS workshop on emerging technologies: wireless power transfer (WoW), Knoxville, TN, 2016, pp 31–34
5. Wang C-S, Covic GA, Stielau OH (2004) Power transfer capability and bifurcation phenomena of loosely coupled inductive power transfer systems. IEEE Trans Ind Electron 51(1):148–157

# Part II
# Multi-resonator WPT Systems

The conventional two-resonator system has two major limitations, i.e., the power transfer distance and the number of the supplying loads. Multi-resonator system is a promising solution to these limitations. This part firstly introduces domino-resonator WPT systems with at least one relay resonators which can prolong the power transfer distance substantially. Later, the studies are extended to another two three-resonator systems in which either two receivers are enabled or the relay resonator is not placed in the middle of the power transfer path.

# Chapter 3
# General Model of Multi-resonator Systems

## 3.1 Circuit Model [1]

Figure 3.1 shows the lumped circuit model for multi-resonator WPT system. It should be noted that the practical resonators could have arbitrary relative positions. Regardless of their positions, the coupling between every two of them can be represented as the mutual inductance. Therefore, for each resonator, one KVL equation can be obtained and totally there are $n$ KVL equations which can be written in matrix form as shown in (3.1).

$$\begin{bmatrix} R_1 + R_{L1} + jX_1 & j\omega M_{12} & j\omega M_{13} & \cdots & \cdots & j\omega M_{1n} \\ j\omega M_{12} & R_2 + R_{L2} + jX_2 & j\omega M_{23} & \cdots & \cdots & j\omega M_{2n} \\ \vdots & \vdots & \ddots & \vdots & \vdots & \vdots \\ j\omega M_{1(n-1)} & \cdots & \cdots & \cdots & R_{n-1} + R_{L(n-1)} + jX_{n-1} & j\omega M_{(n-1)n} \\ j\omega M_{1n} & \cdots & \cdots & \cdots & j\omega M_{(n-1)n} & R_n + R_{Ln} + jX_n \end{bmatrix} \cdot \begin{bmatrix} \mathbf{I_1} \\ \mathbf{I_2} \\ \vdots \\ \mathbf{I_{n-1}} \\ \mathbf{I_n} \end{bmatrix} = \begin{bmatrix} \mathbf{V_{S1}} \\ \mathbf{V_{S2}} \\ \vdots \\ \mathbf{V_{S(n-1)}} \\ \mathbf{V_{Sn}} \end{bmatrix}, \quad (3.1)$$

where, as shown in, $R_u$ is the parasitic resistance (the sum of the resistance of Winding-$u$, the ESR of Capacitor-$u$ and the internal series resistance of Source-$u$), $R_{Lu}$ is the load resistance, $X_u$ is the reactance $\omega L_u - 1/(\omega C_u)$, $L_u$ is the inductance, $C_u$ is the capacitance, $I_u$ is the current and $V_{Su}$ is the voltage source in Resonator-$u$ ($u = 1, 2, \ldots, n$); $M_{ij}$ is the mutual inductance between Resonator-$i$ and Resonator-$j$; $\omega$ is the angular frequency. For a resonator without a load, $R_{Lu} = 0$.


W. Zhong et al., *Wireless Power Transfer*, CPSS Power Electronics Series,
https://doi.org/10.1007/978-981-15-2441-7_3

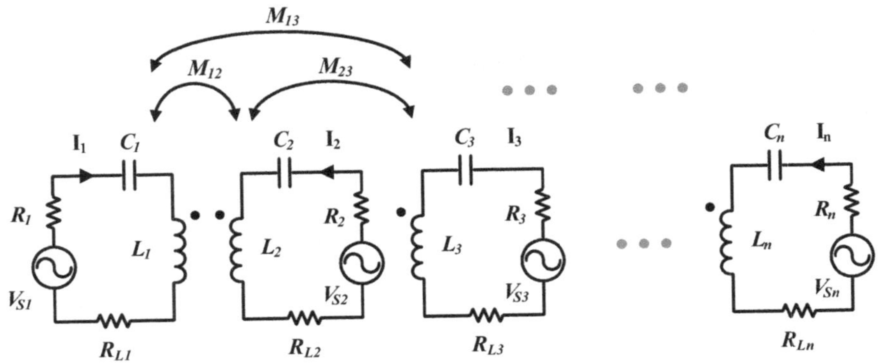


**Fig. 3.1** Lumped circuit model of a multi-resonator system with $n$ resonators

## 3.2 Mutual Inductance Calculation

### 3.2.1 *Coaxial Coils*

For two coaxial circular filamentary current loops, Maxwell [2] has derived the following equation to calculate the mutual inductance:

$$M = \mu_0 \frac{\sqrt{r_1 r_2}}{\alpha}[(2 - \alpha^2)K(\alpha) - 2E(\alpha)], \tag{3.2}$$

where $K(\alpha)$ and $E(\alpha)$ are the complete elliptic integrals of the first and second kind, respectively, and

$$\alpha = \sqrt{\frac{4r_1 r_2}{d^2 + (r_1 + r_2)^2}}, \tag{3.3}$$

where $r_1$, $r_2$ and $d$ are the radii of loop-1, loop-2 and the distance between them, respectively.

For two coaxial circular thin-wall windings, if the dimension of the wire of the windings is relatively small compared with the dimension of the windings so that each turn of the windings can be considered as a filamentary current loop, then the mutual inductance between the two windings can be calculated by:

$$M = \sum_{i=1}^{n_1} \sum_{j=1}^{n_2} M_{ij}, \tag{3.4}$$

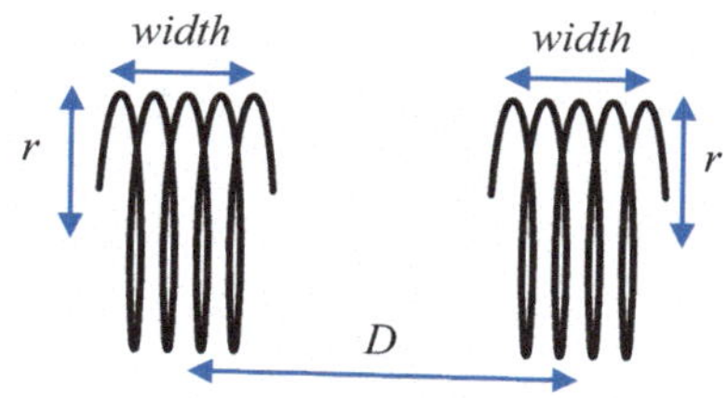


**Fig. 3.2** Two identical helical windings

where $n_1$ and $n_2$ are the numbers of turns of the two windings; $M_{ij}$ is the mutual inductance between the *i*th turn of the first winding and the *j*th turn of the second winding which can be worked out with (3.2).

**Example**

To calculate the mutual inductance of two identical helical windings placed coaxially as shown in Fig. 3.2. Use $D$ to represent the distance between the centers of the windings. The radius, width and number of turns of the windings are 0.155 m, 0.015 m, and 11, respectively. The Matlab program to calculate the mutual inductance is given as follows.

- Matlab Program

```
function M=Mutual_Inductance(D)
r=0.155;
width=0.015;
N=11;
u0=4e-7*pi;
M=0;
for v=1:1:N
    for j=1:1:N
        dis=D+(j-v)*width/11;
        f=4*r^2/(4*r^2+dis^2);
        [k,e] = ellipke(f);
        M=M+u0*r*((2-f)*k-2*e)/f^0.5;
    end
end
```

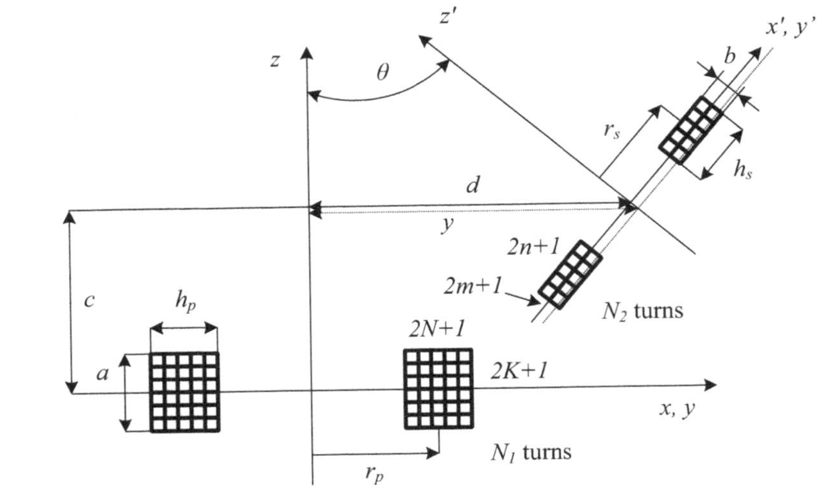


**Fig. 3.3** Diagram of two non-coaxial circular coils [4]

### 3.2.2 *Non-coaxial Coils*

Based on the calculation method for the mutual inductance between two filamentary circular current loops with inclined axes in [3], a mutual inductance equation for non-coaxial coils is proposed in [4] (see Fig. 3.3) as

$$M = \frac{N_1 N_2 \sum_{g=-K}^{K} \sum_{h=-N}^{N} \sum_{l=-n}^{n} \sum_{p=-m}^{m} M(g,h,l,p)}{(2K+1)(2N+1)(2n+1)(2m+1)}, \tag{3.5}$$

where

$$M(g,h,l,p) = \frac{\mu_0}{\pi}\sqrt{r_p(h) r_s(l)} \int_0^{\pi} \frac{\left(\cos\theta - \frac{y(p)}{r_s(l)}\cos\phi\right)\Psi(k)}{V^{3/2}} \mathrm{d}\phi$$
$$V = \sqrt{1 - \cos^2\phi \sin^2\theta - 2\frac{y(p)}{r_s}\cos\phi\cos\theta + \frac{y^2(p)}{r_s^2}}$$
$$\Psi(k) = \left(\frac{2}{k} - k\right)K(k) - \frac{2}{k}E(k) = Q_{1/2}(x),\ x = \frac{2-k^2}{k^2}$$
$$y(p) = d + \frac{b\sin\theta}{2m+1}p,\ p = -m, \ldots, 0, \ldots, m$$
$$r_p(h) = r_p + \frac{h_p}{2N+1}h,\ h = -N, \ldots, 0, \ldots, N$$
$$r_s(l) = r_s + \frac{h_s}{2n+1}l,\ l = -n, \ldots, 0, \ldots, n$$
$$z(g,p) = c + \frac{a}{2K+1}g - \frac{b\cos\theta}{2m+1}p,\ g = -K, \ldots, 0, \ldots, K,\ p = -m, \ldots, 0, \ldots, m$$
$$k^2 = \frac{4\alpha V}{(1+\alpha V)^2 + \xi^2},\ \xi = \beta - \alpha\cos\phi\sin\theta,\ \alpha = \frac{r_s}{r_p(h)},\ \beta = \frac{z(g,p)}{r_p(h)}$$

| | |
|---|---|
| $N_1$ | number of turns of the larger coil; |
| $N_2$ | number of turns of the smaller coil; |
| $r_P$ | radius of the larger coil; |

| | |
|---|---|
| $h_p$ | thickness of the larger coil; |
| $a$ | axial length of the larger coil; |
| $r_s$ | radius of the smaller coil; |
| $h_s$ | thickness of the smaller coil; |
| $b$ | axial length of the smaller coil; |
| $c$ | distance between coils' centers; |
| $d$ | distance between coil planes; |
| $\theta$ | angle between axes; |
| $K(k)$ | complete elliptic integral of the first kind; |
| $E(k)$ | complete elliptic integral of the second kind; |
| $Q_{1/2}(k)$ | Legendre function of the second kind and half-integral degree. |

## 3.3 Efficiency Optimization Methodology [1]

As mentioned in Chap. 2, power transfer efficiency is normally the primary concern in a WPT system unless the transmitted power is very low, for instance, lower than 1 W. As defined previously, the power transfer efficiency ($\eta$) is the ratio between the total output power of the loads over the total input power of the sources:

$$\begin{aligned}\eta &= \frac{I_1^2 R_{L1} + I_2^2 R_{L2} + \cdots + I_n^2 R_n}{I_1^2(R_1 + R_{L1}) + I_2^2(R_2 + R_{L2}) + \cdots + I_n^2(R_n + R_{Ln})} \\ &= \frac{R_{L1} + \frac{I_2^2}{I_1^2} R_{L2} + \cdots + \frac{I_n^2}{I_1^2} R_{Ln}}{R_1 + R_{L1} + \frac{I_2^2}{I_1^2}(R_2 + R_{L2}) + \cdots + \frac{I_n^2}{I_1^2}(R_n + R_{Ln})}.\end{aligned} \tag{3.6}$$

In many application scenarios, a WPT system will contain only one power source. So the following discussion will focus on one-source systems. When the system contains only one source, say $\mathbf{V_{S1}}$, which means $\mathbf{V_{S2}} - \mathbf{V_{Sn}}$ in (3.1) equal to zero, then, all the current ratios (totally $n - 1$ unknowns) in (3.6) can be obtained by solving the KVL equations in all resonators except the transmitter:

$$\begin{bmatrix} j\omega M_{12} & R_2 + R_{L2} + jX_2 & j\omega M_{23} & \cdots & \cdots & j\omega M_{2n} \\ \vdots & \vdots & \vdots & \vdots & \vdots & \vdots \\ j\omega M_{1(n-1)} & \cdots & \cdots & \cdots & R_{n-1} + R_{L(n-1)} + jX_{n-1} & j\omega M_{(n-1)n} \\ j\omega M_{1n} & \cdots & \cdots & \cdots & j\omega M_{(n-1)n} & R_n + R_{Ln} + jX_n \end{bmatrix} \cdot \begin{bmatrix} \mathbf{I_1} \\ \vdots \\ \mathbf{I_{n-1}} \\ \mathbf{I_n} \end{bmatrix} = \mathbf{0}. \tag{3.7}$$

As can be implied from (3.6) and (3.7), the compensation condition in the transmitter, i.e., $X_1$, will not affect the efficiency of the system. It only affects the total input impedance of the system, which is similar to the case in 2-resonator systems.

If the construction of a WPT system is given, or in other words, the known parameters in (3.7) are $L_i$ ($i = 1, 2, \ldots, n$) and $M_{ij}$ ($i = 1, 2, \ldots, n - 1; j = i + 1, i + 2, \ldots, n$); while the unknowns are $C_i$ ($i = 1, 2, \ldots, n$), the operating frequency $f$ and the load resistances. The power source is assumed to be ideal which means the source resistance is zero. Generally, a multi-resonator system can be optimized under following different scenarios:

1. For a multi-resonator system with given resonators (i.e. $L_i$ and $C_i$ fixed), a given structure (i.e. the mutual inductance $M_{ij}$ fixed) and a given operating frequency (i.e. $f$ and the resistance $R_i$ fixed), the efficiency of the system is a function of the load resistance:

$$\eta = \mathrm{f}(R_{L1}, R_{L2}, \ldots, R_{Ln}). \tag{3.8}$$

2. For a system with given resonators (i.e. $L_i$ and $C_i$ fixed), a given structure (i.e. mutual inductance $M_{ij}$ fixed), and assuming that the operating frequency varies in a small range so that the resistance $R_i$ can be considered unchanged, then the efficiency of the system is a function of the operating frequency and the load resistances:

$$\eta = \mathrm{f}(f, R_{L1}, R_{L2}, \ldots, R_{Ln}). \tag{3.9}$$

3. For a system with given coils (i.e. $L_i$ fixed), a given structure (i.e. mutual inductance $M_{ij}$ fixed), and a given operating frequency (i.e. $f$ and $R_i$ fixed), the efficiency of the system is a function of the compensating capacitances and the load resistances:

$$\eta = \mathrm{f}(C_2, \ldots, C_n, R_{L1}, R_{L2}, \ldots, R_{Ln}). \tag{3.10}$$

4. For a system with given coils (i.e. $L_i$ fixed) and a given structure (i.e. mutual inductance $M_{ij}$ fixed), the efficiency of the system can be expressed as:

$$\eta = \mathrm{f}(f, C_1, \ldots, C_n, R_{L1}, R_{L2}, \ldots, R_{Ln}). \tag{3.11}$$

In some applications, the positions of the resonators might be adjustable. Then the mutual inductances are dependent on the relative positions between every two coils.

Therefore, the relative distances and angles could also be considered as variables and optimized in the these applications.

After the efficiency function of a system is obtained with the help of (3.7), the optimization could be done by using Matlab Optimization Toolbox [5].

## References

1. Zhong W, Lee CK, Ron Hui SY (2013) General analysis on the use of Tesla's resonators in domino forms for wireless power transfer. IEEE Trans Ind Electron 60(1):261–270
2. Maxwell JC (1873) A treatise on electricity and magnetism. Dover, New York (1954) (reprint from the original from 1873)
3. Grover FW (1964) Inductance calculations. Dover, New York
4. Babic SI, Akyel C (2008) Calculating mutual inductance between circular coils with inclined axes in air. IEEE Trans Magn 44(7):1743–1750
5. [Online] https://www.mathworks.com/products/optimization.html

# Chapter 4
# Straight Domino-Resonator Systems

## 4.1 Introduction

As shown in Fig. 2.5, the power transfer efficiency of a conventional 2-resonator system decays exponentially as the power transfer distance increases, therefore, the power transfer distance of a 2-resonator system is typically less than 2–3 times the diameter of the coils. References [1, 2] proposed a domino-resonator structure to substantially increase the power transfer distance of a WPT system. In this structure, relay resonators are added in between the transmitting resonator and the receiving resonator. A straight domino system is shown in Fig. 4.1. This chapter presents studies on the following three aspects of a straight domino system.

1. The efficiency decrease rate of a straight domino system with respect to the power transfer distance. Based on a case study, it is shown that the efficiency of a domino system with a fixed distance between every two adjacent resonators decreases linearly as the power transfer distance increases, which is much slower than that of a conventional 2-resonator system.
2. The effects of the coupling between non-adjacent coils to the power flow and efficiency of a domino system are addressed. A methodology for power flow analysis is introduced. Based on this, it is proved that the optimal operating frequency of a straight domino system is no longer the natural resonant frequency of the resonators as in a conventional 2-resonator system.
3. For a given total power transfer distance, the spacing of the resonators are optimized for a maximum efficiency. In waveguide applications, straight arrangements of resonators are also used and the resonators are equally spaced to achieve impedance matching and maximum power transmission [3]. However, for power applications, power losses of the system become a major concern. So the effect of the spacing of the resonators on the energy efficiency is investigated. It turns out that the optimal arrangement of the resonators in a straight domino pattern is not necessarily equal-spacing to achieve a maximum efficiency.


W. Zhong et al., *Wireless Power Transfer*, CPSS Power Electronics Series,
https://doi.org/10.1007/978-981-15-2441-7_4

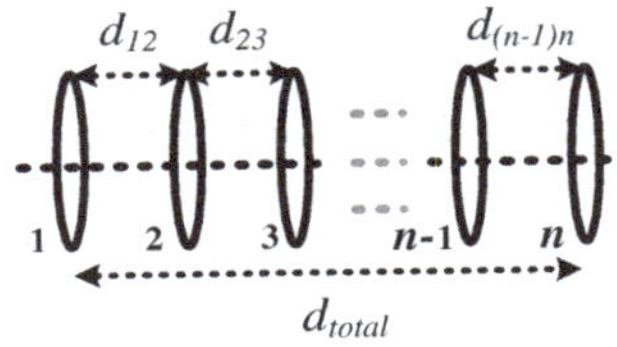


**Fig. 4.1** Schematic of a straight domino-resonator system with $n$ resonators (Source is connected to Resonator-1 and load is connected to Resonator-$n$.)

## 4.2 Efficiency of a Straight Domino-Resonator System [1]

Practical resonators, as shown in Fig. 4.2, have been built and used in case studies. The parameters of the resonators are listed in Table 4.1. Calculations are carried out assuming the distance between every two adjacent resonators of the domino systems are fixed at 0.3 m. For example, when the power transfer distance is 2.1 m, totally 8 resonators are used. The efficiency of the domino system can be worked out as a function of the load resistance in Matlab based on (3.6) and (3.7). Then the Matlab

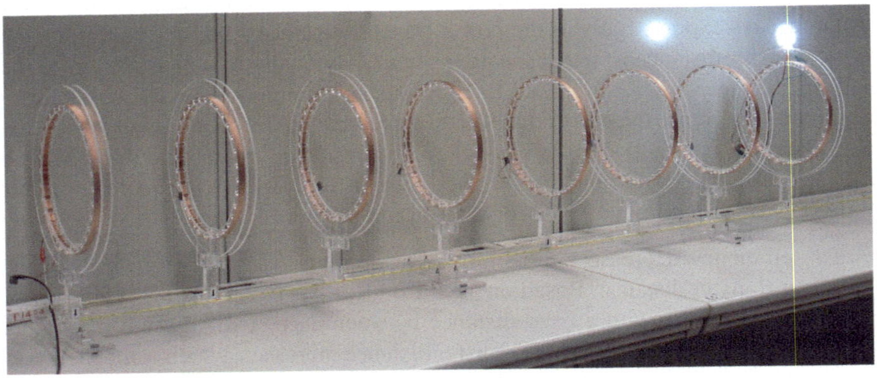

**Fig. 4.2** A practical straight domino-resonator system (The first resonator in the left is connected to a high-frequency AC power source.)

**Table 4.1** Parameters of the practical resonators

| | | | |
|---|---|---|---|
| Radius of the coil | 155 mm | Layers of the wire | 1 |
| Number of turns | 11 | Axial Length of the coil | 15 mm |
| Structure of the wire | Ø0.12 mm × 50 strands<br>Outer Ø1.2 mm | Inductance | 90.7 μH |
| Capacitance | 1.036 nF<br>(Sum of the external capacitance 1.019 nF and the parasitic capacitance of the coil 0.017 nF)<br>**[resonant frequency is 519.2 kHz]** | Resistance<br>(at 520 kHz) | 0.92 Ω |

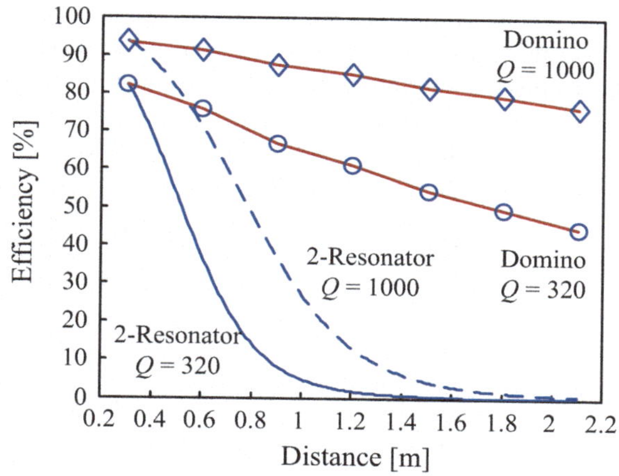


**Fig. 4.3** Maximum efficiency comparison between a two-resonator system and an equal-spacing straight domino-resonator system

optimization function **fminsearch** could be used to obtain the maximum efficiency of the system. The maximum efficiency of the domino system with $Q$ factors of 1000 (an assumed value) and 320 (the practical value) are calculated and plotted in Fig. 4.3. It can be seen that the efficiencies of the domino systems decrease linearly with power transfer distance. Based on (2.25), the maximum efficiencies of a 2-resonator system with the same $Q$ factors are also included and they exhibit exponential decay as expected.

## 4.3 Methodology for Power Flow Analysis [2]

One feature of the domino-resonator system is that there are cross-couplings between non-adjacent resonators. In this section, a power flow analysis that identifies the power flow paths and their equations is presented.

The methodology of the power flow analysis is demonstrated with the use of a 3-resonator system as shown in Fig. 4.4. In this example, there are two power paths from the source to the load. The first path involves adjacent resonators 1 and 2 and adjacent resonators 2 and 3 (i.e. from 1 to 2 to 3). The other path involves non-adjacent resonators 1 and 3 (i.e. from 1 to 3), namely the cross-coupling path.

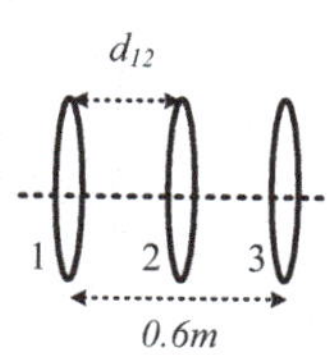


**Fig. 4.4** A straight 3-resonator system with total transfer distance of 0.6 m (*Note* Resonator-1 is the transmitter and Resonator-3 is the receiver.)

The power flow in each path can be worked out in order to understand how the power flows will change under frequency variation. First, the current ratios can be determined by solving (3.7) for the 3-resonator system. The results come out as:

$$\frac{\mathbf{I}_2}{\mathbf{I}_1} = \frac{-\omega M_{12} R_{\text{out}} + j(\omega^2 M_{13} M_{23} - \omega M_{12} X_3)}{R_2 X_3 + R_{\text{out}} X_2 + j(X_2 X_3 - R_2 R_{\text{out}} - \omega^2 M_{23}^2)} \tag{4.1}$$

$$\frac{\mathbf{I}_3}{\mathbf{I}_1} = \frac{-\omega M_{13} R_2 + j(\omega^2 M_{12} M_{23} - \omega M_{13} X_2)}{R_2 X_3 + R_{\text{out}} X_2 + j(X_2 X_3 - R_2 R_{\text{out}} - \omega^2 M_{23}^2)}, \tag{4.2}$$

where

$$\begin{aligned} X_2 &= \omega L_2 - 1/(\omega C_2); \\ X_3 &= \omega L_3 - 1/(\omega C_3); \\ R_{\text{out}} &= R_3 + R_L. \end{aligned}$$

Now assuming the input current is 1 A, with the help of (4.1) and (4.2), the complex power flows can be expressed as

$$\begin{aligned} \mathbf{S}_{12} &= P_{12} + jQ_{12} = -j\omega M_{12}\mathbf{I}_1\mathbf{I}_2^* \\ &= \frac{\left(\begin{pmatrix} \omega^2 M_{12}^2 X_3 R_2 X_3 + \omega^2 M_{12}^2 R_{\text{out}}^2 R_2 + \omega^4 M_{12}^2 M_{23}^2 R_{\text{out}} \\ - \omega^3 M_{12} M_{13} M_{23} R_2 X_3 - \omega^3 M_{12} M_{13} M_{23} R_{\text{out}} X_2 \end{pmatrix} + j\begin{pmatrix} \omega^3 M_{12} M_{13} M_{23} X_2 X_3 + \omega^4 M_{12}^2 M_{23}^2 X_3 - \omega^5 M_{12} M_{13} M_{23}^3 \\ - \omega^3 M_{12} M_{13} M_{23} R_2 R_{\text{out}} - \omega^2 M_{12}^2 R_{\text{out}}^2 X_2 - \omega^2 M_{12}^2 X_3 X_2 X_3 \end{pmatrix}\right)}{(R_2 X_3 + R_{\text{out}} X_2)^2 + (X_2 X_3 - R_2 R_{\text{out}} - \omega^2 M_{23})^2} \end{aligned} \tag{4.3}$$

$$\begin{aligned} \mathbf{S}_{23} &= P_{23} + jQ_{23} = -j\omega M_{23}\mathbf{I}_2\mathbf{I}_3^* \\ &= \frac{\left(\begin{pmatrix} \omega^3 M_{12} M_{23} M_{13} R_2 X_3 + \omega^4 M_{12}^2 M_{23}^2 R_{\text{out}} \\ - \omega^3 M_{12} M_{23} M_{13} R_{\text{out}} X_2 - \omega^4 M_{13}^2 M_{23}^2 R_2 \end{pmatrix} + j\begin{pmatrix} \omega^5 M_{12} M_{23}^3 M_{13} + \omega^3 M_{12} M_{23} M_{13} X_2 X_3 \\ - \omega^3 M_{12} M_{23} M_{13} R_{\text{out}} R_2 - \omega^4 M_{12}^2 M_{23}^2 X_3 - \omega^4 M_{13} M_{23}^2 M_{13} X_2 \end{pmatrix}\right)}{(R_2 X_3 + R_{\text{out}} X_2)^2 + (X_2 X_3 - R_2 R_{\text{out}} - \omega^2 M_{23}^2)^2} \end{aligned} \tag{4.4}$$

$$\begin{aligned} \mathbf{S}_{13} &= P_{13} + jQ_{13} = -j\omega M_{13}\mathbf{I}_1\mathbf{I}_3^* \\ &= \frac{\left(\begin{pmatrix} \omega^2 M_{13}^2 R_2^2 R_{\text{out}} + \omega^4 M_{23}^2 M_{13}^2 R_2 + \omega^2 M_{13}^2 R_{\text{out}} X_2^2 \\ - \omega^3 M_{12} M_{23} M_{13} R_2 X_3 - \omega^3 M_{12} M_{23} M_{13} R_{\text{out}} X_2 \end{pmatrix} + j\begin{pmatrix} \omega^2 M_{13}^2 R_2^2 X_3 + \omega^2 M_{13}^2 X_2^2 X_3 + \omega^3 M_{12} M_{23} M_{13} R_2 R_{\text{out}} \\ + \omega^5 M_{12} M_{23}^3 M_{13} - \omega^3 M_{12} M_{23} M_{13} X_2 X_3 - \omega^4 M_{23}^2 M_{13}^2 X_2 \end{pmatrix}\right)}{(R_2 X_3 + R_{\text{out}} X_2)^2 + (X_2 X_3 - R_2 R_{\text{out}} - \omega^2 M_{23}^2)^2} \end{aligned} \tag{4.5}$$

$$\begin{aligned}\mathbf{S}_3 &= P_3 + jQ_3 = |\mathbf{I}_3|^2(R_{\text{out}} + jX_3)\\ &= \frac{\omega^2 M_{13}^2 R_2^2 + (\omega^2 M_{12}M_{23} - \omega M_{13}X_2)^2}{(R_2X_3 + R_{\text{out}}X_2)^2 + (X_2X_3 - R_2R_{\text{out}} - \omega^2 M_{23}^2)^2}(R_{\text{out}} + jX_3),\\ &= \mathbf{S}_{23} + \mathbf{S}_{13}\end{aligned} \tag{4.6}$$

where $\mathbf{S}_{ij}(i, j = 1, 2, 3)$ represents the complex power generated in Resonator-$j$ due to the current in Resonator-$i$ and $\mathbf{S}_3$ represents the total complex power in Resonator-3.

## 4.4 Effects of Cross-Coupling [2]

To investigate the effects of the cross-couplings of non-adjacent resonators, an equally-spaced 3-resonator system is used as an example. The coils are the same as those used in the last section (Table 4.1). In the system, the three resonators are equally spaced with a total transfer distance of 0.6 m and a load resistance of 10 Ω.

In order to highlight the effects of the cross-coupling, the individual power flow curves and the overall efficiency of the system without and with the inclusion of the cross-coupling are plotted as a function of the operating frequency in Fig. 4.5 and Fig. 4.6, respectively. The natural resonant frequency of the resonator (i.e. $f_0 =$ 519.2 kHz) is chosen as the base value for the operating frequency. It can be seen that:

(1) Without considering the cross-coupling of Resonators-1 and Resonators-3, the optimal operating frequency got from the numerical optimization method is 519.7 kHz. Therefore, the resonant frequency can be considered as the optimal

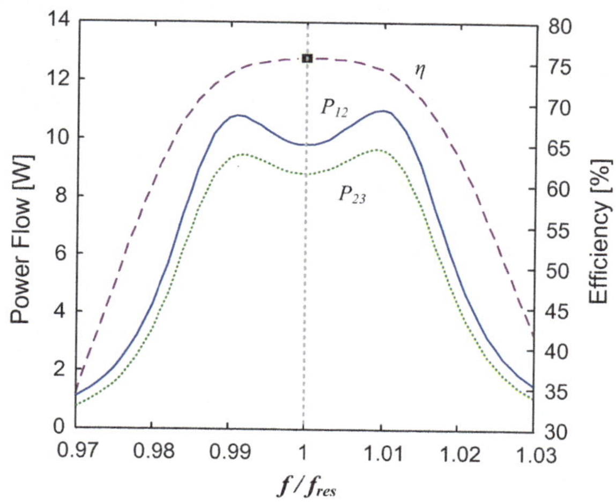


**Fig. 4.5** Variations of the power flow and the efficiency with the operating frequency of an equally-spaced 3-resonator system in a straight domino arrangement, ***excluding*** the effects of the cross-coupling of the non-adjacent resonators (*Note* Frequency is normalized by the resonant frequency.)

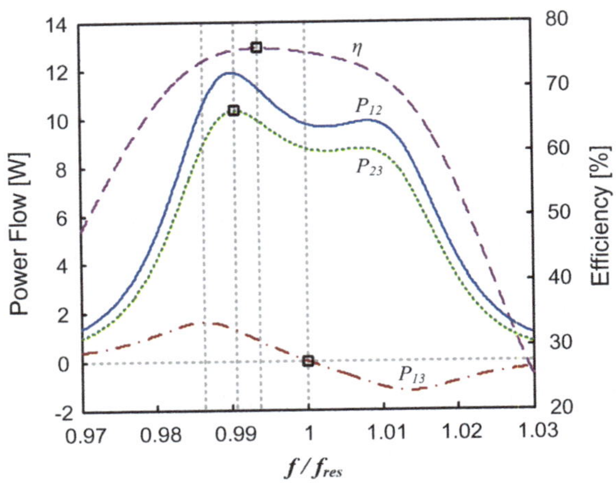


**Fig. 4.6** Variations of the power flow and the efficiency with the operating frequency of an equally-spaced 3-resonator system in a straight domino arrangement, ***including*** the effects of the cross-coupling of the non-adjacent resonators

operating frequency when the cross-couplings are neglected in a straight domino system.

(2) With the cross-coupling included, it can be seen from Fig. 4.6 that the shapes of the power flow curves of $P_{12}$, $P_{23}$ are very different from those in Fig. 4.5. $P_{13}$ is close to zero at the resonant frequency, positive below the resonant frequency and negative above the resonant frequency. The maximum points of $P_{12}$ and $P_{23}$ occur below the resonant frequency.

(3) The consequence of the inclusion of the cross-coupling is that the maximum efficiency operation occurs below the resonant frequency in this 3-resonator system which can be observed from Fig. 4.6.

The analysis bellow provides some physical insights into these observations. At the resonant frequency, the cross-coupling power flow $P_{13}$ in the path 1–3 approaches zero. The reason can be deduced from the following equations of the system operating at the resonant frequency:

$$j\omega M_{12}\mathbf{I_1} + R_2\mathbf{I_2} + j\omega M_{23}\mathbf{I_3} = 0 \tag{4.7}$$

$$j\omega M_{13}\mathbf{I}_1 + j\omega M_{23}\mathbf{I_2} + (R_3 + R_L)\mathbf{I_3} = 0. \tag{4.8}$$

From (4.7) and (4.8),

$$\mathbf{I_3} = -\frac{\omega^2 M_{12}M_{23} + j\omega M_{13}R_2}{\omega^2 M_{23}^2 + R_2(R_3 + R_L)}\mathbf{I_1}. \tag{4.9}$$

In the equally-spaced three-resonator system described above, according to (3.2) and (3.4), $M_{12} = M_{23} = 2.857 \times 10^{-6}$ H, and $M_{13} = 5.326 \times 10^{-7}$ H, thereby,

$$\omega^2 M_{12} M_{23} / \omega M_{13} R_2 = 55.55. \tag{4.10}$$

So $\omega^2 M_{12} M_{23} \gg \omega M_{13} R_2$, and thus the imaginary part of the numerator in (4.9) can be neglected, and we have

$$\mathbf{I}_3 \approx -\frac{\omega^2 M_{12}^2}{\omega^2 M_{12}^2 + (R_3 + R_L) R_2} \mathbf{I}_1. \tag{4.11}$$

Therefore, the currents in Resonator-3 and 1 are essentially out of phase and therefore the complex power $(j M_{13} \mathbf{I}_1) \cdot \mathbf{I}_3^*$ delivered from Resonator-1 to Resonator-3 directly contains reactive power only. This deduction is valid for the systems in which the distance between non-adjacent resonators is relatively large.

The maximum efficiency of the system will be achieved only when both of the power flow paths (i.e. the path via the adjacent resonators 1–2–3 and the path via the non-adjacent resonators 1–3) are optimally utilized. In Fig. 4.6, the four vertical dotted lines indicate four special frequencies which are, from right to left, the resonant frequency and the frequencies at which the maximum values of $\eta$, $P_{23}$ and $P_{13}$, respectively, occur. Theoretical values at these four frequencies are tabulated in Table 4.2. At the resonant frequency, the power flow via the non-adjacent resonators ($P_{13}$) is virtually zero, meaning that all the power must flow through the relay resonator (i.e. the adjacent resonator path 1–2–3). Therefore, it is not the most efficient operation because the cross-coupling path is **not** utilized. In this case, the cross-coupling path 1–3 has relatively small mutual inductance. Nevertheless, it can still provide positive power flow to increase the overall performance of the system when the operating frequency is slightly below the resonant frequency.

**Table 4.2** Power flow of the 3-resonator system under different frequencies

| $f/f_{res}$ | 0.9864 | 0.9906 | 0.9937 | 1 |
|---|---|---|---|---|
| $\eta$ *(purple curve)* (%) | 73.99 | 75.70 | **75.99** | 75.23 |
| $P_{12}$ (W) | 10.61 | 11.92 | 11.27 | 9.794 |
| $P_{23}$ (W) | 8.97 | 10.35 | 9.936 | 8.766 |
| $P_{13}$ (W) | 1.626 | 1.299 | 0.8421 | 0.0281 |
| $\eta_2$ (%) ($P_{23}/P_{12}$) | 84.54 | 86.83 | 88.16 | 89.50 |

## 4.5 Spacing Optimization of a Straight Domino-Resonator System [2]

### 4.5.1 Three-Resonator System

Besides of the operating frequency, the spacing could be optimized in a straight domino system. Based on the same 3-resonator system described in the last section, Fig. 4.7 shows the efficiency curve of the system with both the operating frequency and distances among resonators optimized. The efficiency curve at the resonant frequency is included in Fig. 4.7 for comparison. It can be seen that the efficiency can be improved over a wider range of load resistance if both of the operating frequency and spacing are optimized. The normalized optimal frequency curve ($f_{opt}/f_{res}$) is also plotted in Fig. 4.7. It is noted that the optimal frequency can be higher or lower than the resonant frequency depending on the load resistance.

For this 3-resonator system with an overall transfer distance of 0.6 m, the variation of the optimal spacing between the 1st and the 2nd resonators with a wide range of load resistance is plotted in Fig. 4.8. If maximum efficiency is set as the objective, it is noted that the relay resonator (i.e. Resonator-2) should be placed closer to Resonator-1 for a smaller $R_L$ and closer to Resonator-3 when $R_L$ becomes larger.

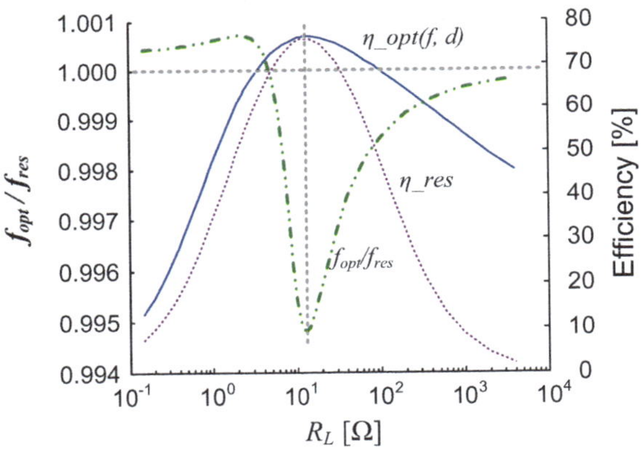


**Fig. 4.7** Optimal frequency and efficiency curves of the 3-resonator system for a wide range of load resistance (η_opt(*f, d*) is the efficiency curve of the system with the operating frequency and the spacing optimized, and η_res is efficiency curve of the equally-spaced system operating at the resonant frequency. $f_{opt}/f_{res}$ represents the curve of the optimal frequency normalized by the resonant frequency.)

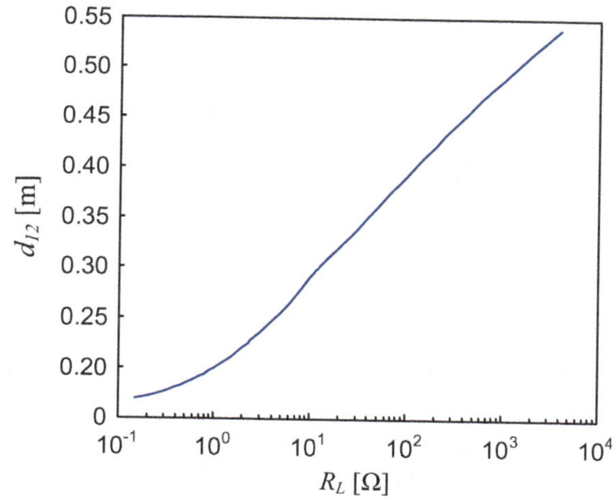


**Fig. 4.8** Variation of the optimal distance between the 1st and 2nd resonators ($d_{12}$) of the 3-resonator system with an overall transfer of 0.6 m for a wide range of load resistance at the optimal operating frequencies

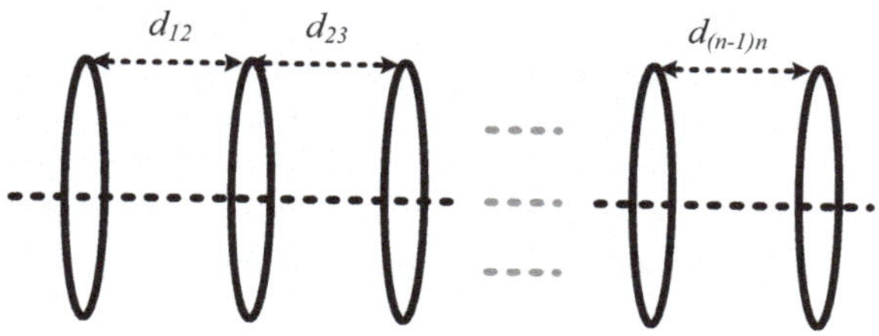


**Fig. 4.9** A general straight domino-system with $n$ resonators

## 4.5.2 n-*Resonator System*

In this part of the analysis, the investigation is extended to an $n$-resonator domino system as shown in Fig. 4.9. Based on a fixed "average distance" $d_e$ between two adjacent resonators, the performance of the systems as a function of the number of resonators is observed. For a straight $n$-resonator system, the total transfer distance $d_{\text{total}}$ equals to $(n-1)d_e$. For example, $d_{\text{total}} = 2d_e$ for a 3-resonator system. The equally- and optimally-spaced systems are studied and compared.

For the equally-spaced systems, the objective is to maximize the efficiency, with the load resistance as an optimized variable, subject to the constraints of equal-spacing of the resonators and operating at the resonant frequency of the resonators as follows:

$$\text{Maximize:}\quad \eta = \text{function}\,(R_L)$$
$$\text{Constraints:}\quad d_{12} = d_{23} = \cdots = d_{(n-1)n} = d_e = d_{\text{total}}/(n-1)$$
$$f = f_0,$$

where $d_{\text{total}}$ is the total transfer distance; $d_e$ is the average distance and $f_0$ is the resonant frequency of the resonators.

For the optimally-spaced systems, the objective is to maximize the efficiency, with the operating frequency, the spacing between adjacent resonators and the load resistance as optimized variables, subject to a certain overall transfer distance as follows:

$$\begin{aligned} \text{Maximize:} \quad & \eta = \text{function}\,(R_L, f, d_{12}, d_{23}, \cdots, d_{(n-1)n}) \\ \text{Constraint:} \quad & d_{12} + d_{23} + \cdots + d_{(n-1)n} = d_{\text{total}}. \end{aligned}$$

The systems with average distances of 0.1, 0.2, 0.3 and 0.4 m are examined for straight domino-resonator systems with the number of resonators ranging from 3 to 8. Figure 4.10 shows the comparison of the efficiency curves based on the two sets of optimization criteria for these four averaged distances. Regardless of the overall transfer distance and the number of resonators, the systems with the operating frequency, spacing and load resistance optimized consistently outperform those with only load resistance optimized and operating at the resonant frequency in terms of efficiency. That is, $\eta$_opt($R_L$, $f$, $d$) is higher than $\eta$_opt($R_L$) in all cases.

The variations of the four optimal frequency curves as a function of the number of resonators are plotted in Fig. 4.11. It is important to note that the optimal frequencies occur below the resonant frequency for their respective optimized load resistances. It is also noted that the deviation from the resonant frequency increases as the averaged distance between resonators decreases and the number of resonators increases. With a smaller overall distance and a larger number of resonators, the cross-coupling paths between non-adjacent resonators become more effective and so the optimal frequency shift further away from the resonant frequency.

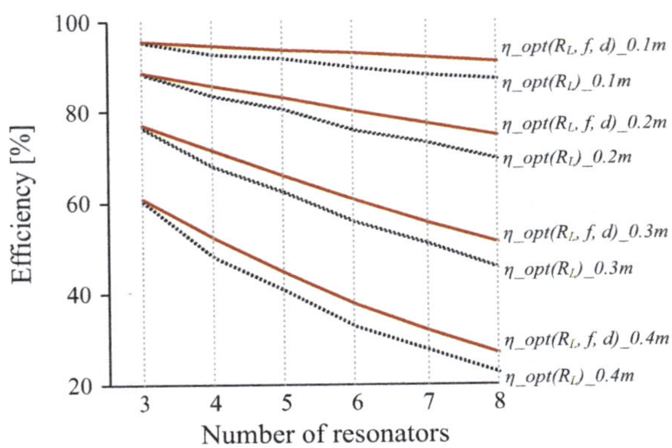


**Fig. 4.10** Efficiency comparison between the domino systems with load resistances, frequencies and distances optimized [$\eta$_opt($R_L$, $f$, $d$)] and the equally spaced systems operating at the resonant frequency and with only load resistance optimized [$\eta$_opt($R_L$)] for four different average distances: 0.1, 0.2, 0.3 and 0.4 m

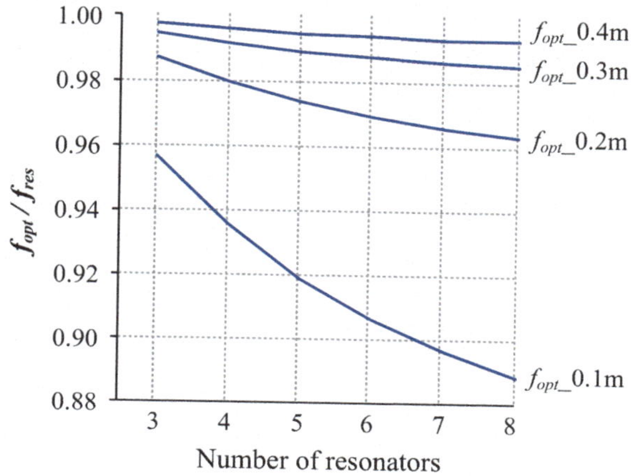


**Fig. 4.11** Optimal frequency normalized by the resonant frequency as a function of the number of resonators

If the overall distance $d_{total}$ between the transmitting resonator (labeled as **T**) and the receiving resonator (labeled as **R**) is normalized as one, the relative positions of the equally-spaced and optimally-spaced relay resonators for the cases of using three to eight resonators are plotted in Fig. 4.12. The real lines indicate the positions for the equally-spaced resonators and the circles indicate the positions for the optimal-spaced resonators. It is noted that the optimal spacing starts to share a similar pattern when the average distances are large enough which makes the width of the coils negligible. One obvious feature of the patterns is that the distance between Resonator-1 and Resonator-2 and that between Resonator-$(n-1)$ and Resonator-$n$ (i.e. $d_{12} > d_{(n-1)n}$) are smaller than average distances (i.e. $d_{12} < d_e$ and $d_{(n-1)n} < d_e$) when $n > 3$.

## 4.6 Summary

When the number of resonators is more than two, both the couplings and the power paths will be more than one. Therefore, the basic operating principle of *setting the operating frequency to the resonant frequency of the coil-resonators* for the two-coil system is no longer valid for multi-resonators systems with more than two coils. Instead, the optimal operating frequency will depends on the couplings (i.e. relative positions of the coils). Moreover, the spacing of a straight domino WPT system can also be optimized.

**Fig. 4.12** Equal and optimal spacing of the resonators in the domino-systems

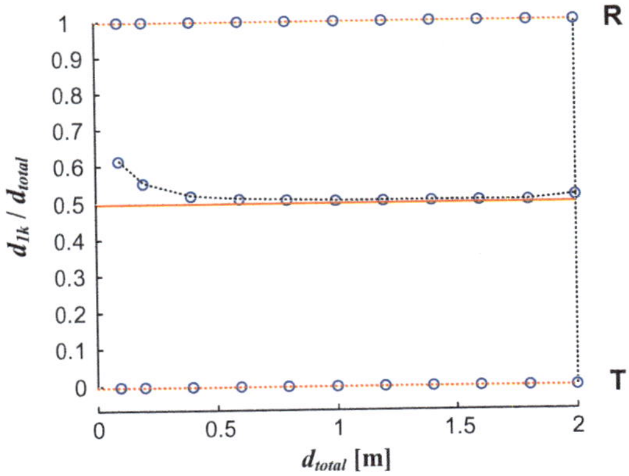


(a) three-resonator

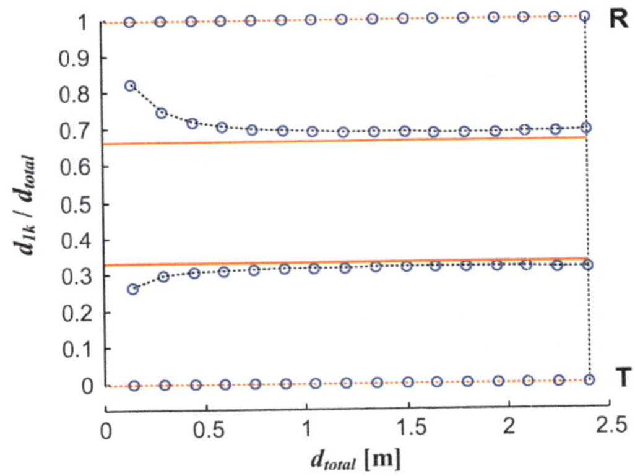


(b) four-resonator

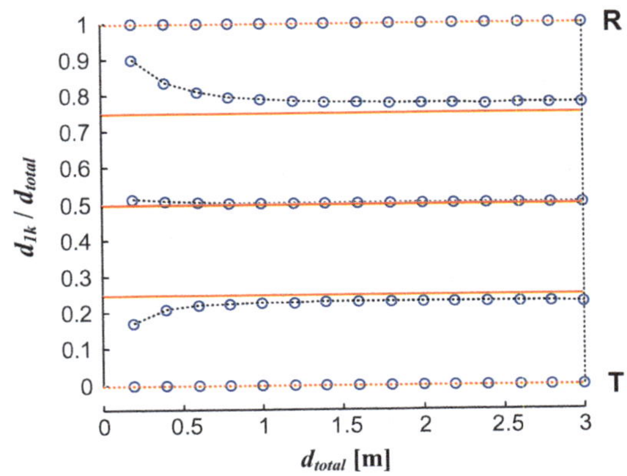


(c) five-resonator

**Fig. 4.12** (continued)

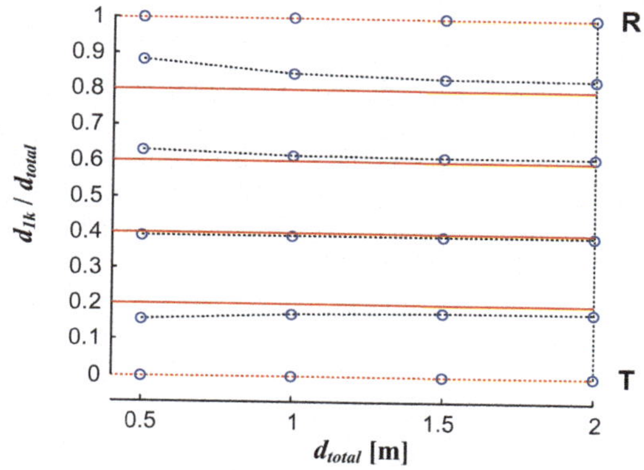


(d) six-resonator

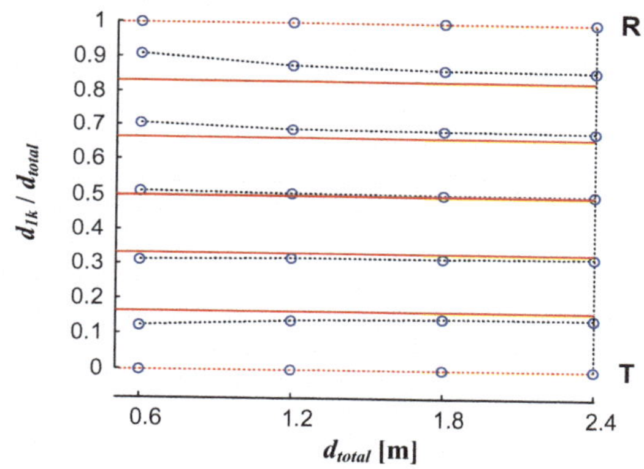


(e) seven-resonator

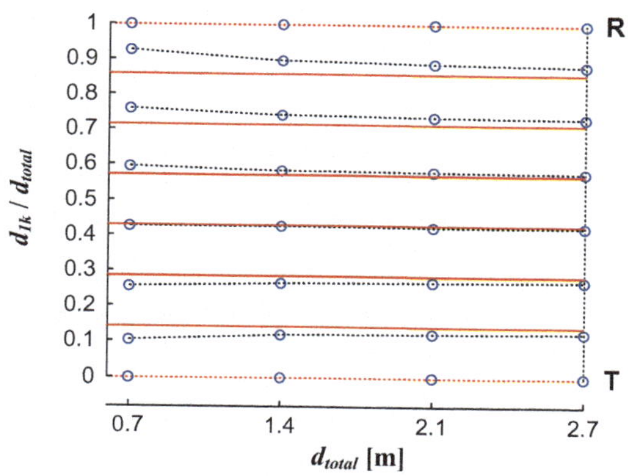


(f) eight-resonator

## References

1. Zhong WX, Lee CK, Hui SYR (2013) General analysis on the use of Tesla's resonators in domino forms for wireless power transfer. IEEE Trans Ind Electron 60(1):261–270
2. Lee CK, Zhong WX, Hui SYR (2012) Effects of magnetic coupling of non-adjacent resonators on wireless power domino-resonator systems. IEEE Trans Power Electron 27(4):1905–1916
3. Shamonina E, Kalinin VA, Ringhofer KH, Solymar L (2002) Magneto-inductive waveguide. Electron Lett 38:371–373

# Chapter 5
# Circular Domino-Resonator Systems

## 5.1 Introduction

A straight domino system constructed with coaxial coils has only one main power path. However, a domino-resonator system may contain non-coaxial resonators [1]. One typical example is a circular domino-resonator system (see Fig. 5.1) which has two main power paths, namely the clockwise and anticlockwise paths. In this chapter, the general analysis of domino-resonator systems with non-coaxial axes is presented. The mathematical formulation of domino-resonator systems with non-coaxial resonators is employed to analyze circular domino systems that have two main power flow paths. The interactions of the two paths are analyzed with superposition principle and explained with the help of vector diagrams. It is discovered that operating at the resonant frequency, current vector cancelation and re-enforcement could occur in some resonators of the systems with even number of resonators. While in systems with odd number of resonators, the currents neither cancel nor re-enforce each other. Therefore, in order to regulate the phases of the load currents from different power paths so that the load currents can re-enforce each other furthest and the maximum efficiency can be achieved, the operating frequency should shift away from the resonant frequency as has explained in a straight domino system in the last chapter or even better, the impedance of each resonator should be adjusted by tuning the compensating capacitance. Moreover, it has been shown that adjusting the compensating capacitances could even offer larger high-efficiency frequency band for domino-resonator systems.

A practical example of the circular domino-resonator system under investigation is shown in Fig. 5.1 in which eight resonators are placed in a circular domino form. These resonators are used in several circular forms in this study and they are the same as those used in Chap. 4. The parameters of the resonators are listed in Table 4.1.



W. Zhong et al., *Wireless Power Transfer*, CPSS Power Electronics Series,
https://doi.org/10.1007/978-981-15-2441-7_5

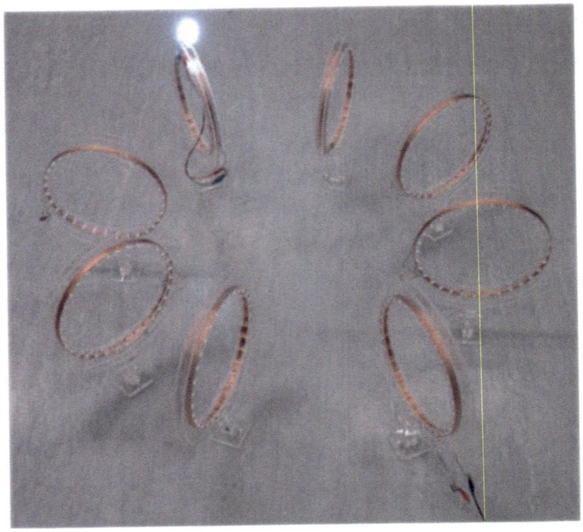

**Fig. 5.1** A practical example of a circular wireless domino-resonator system powering a compact fluorescent lamp (the resonator at the bottom right corner is connected to the power source)

## 5.2 Model of the Circular Domino-Resonator System

With the use of the general circuit equation (3.1) and the mutual inductance equation (3.5), domino-resonator systems with non-coaxial resonators can be analyzed. Figure 5.2 shows the configuration of a circular domino-resonator system with $n$ identical circular resonators in which all the centers of the resonators are placed on a circular path with radius $r_{\text{path}}$ and the center of each resonator is placed in the same plane with the center of the circular path. Other parameters in Fig. 5.2 are: $r_w$ is the radius of the coils; $\theta_1$ to $\theta_n$ are the angles between every two adjacent resonators. For a given set of parameters of the resonators, the number of the resonators ($n$), and the load resistance, the circular domino-resonator system can be defined by the radius of the circular path ($r_{\text{path}}$) and the angles between every two adjacent resonators ($\theta_1$ to $\theta_n$) with the help of (3.5).

In order to get some physical insights into the mutual coupling between two resonators placed along the circular path, two resonators as shown in Fig. 5.3 are used to illustrate the variation of the mutual inductance as a function of the ratio between the radius of the circular path and the radius of the coil. Figure 5.4 shows the theoretical variations of the mutual inductance of two filamentary current loops ($r_w$ of 0.5 m) with the angular displacement ($\theta$) for a range of radii of the circular path ($r_{\text{path}}$). It is noted that the mutual inductance decreases rapidly with increasing $\theta$.

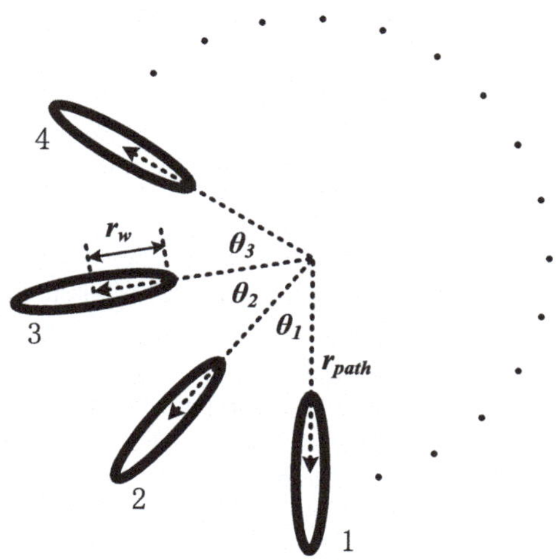


**Fig. 5.2** Configuration of a circular domino-resonator system with $n$ identical coil resonators

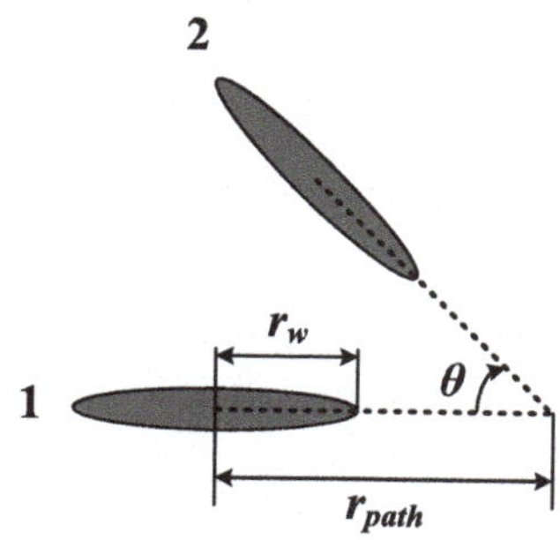


**Fig. 5.3** Two resonators placed in a circular curve

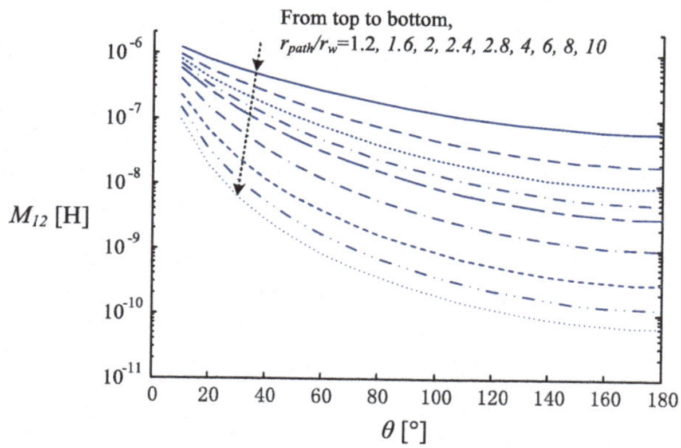


**Fig. 5.4** Variation of the mutual inductance of two filamentary current loops with angular displacement for a set of $r_{path}/r_w$ ratios

## 5.3 Simplified Analysis Without Cross-Couplings

### (i) Assumptions

A simplified analysis on the circular domino-resonator systems is carried out with the following assumptions:

1. In the general circular system of Fig. 5.2, Resonator-1 is excited by an AC power source at the resonant frequency of the resonators and only Resonator-2 to Resonator-$n$ could be loaded.
2. The same resonators as described in Chap. 4 are used. All resonators are evenly placed in the circular path in each setup and test which means the angular displacements are the same ($\theta_1 = \ldots = \theta_n$).
3. All the mutual inductances between two adjacent coils are the same (i.e. 2.647 μH) in all the tested systems in order to compare the performances between the different systems with different numbers of resonator. This means that the radius of the circular path will be changed accordingly when the total number of the resonators in the system is changed.
4. Cross-couplings between nonadjacent resonators are neglected.

### (ii) Efficiency of Circular Domino-Resonator Systems with Different Loaded Positions

For a circular domino-resonator system with all the parameters given except the load resistance, the efficiency of the system will reach the peak when the value of the load resistance $R_L$ is optimized. Based on the lumped circuit model introduced in Chap. 3 and the Matlab optimization tool, the maximum efficiencies of a range of domino-resonator systems with the number of resonators from 3 to 10 have been computed under the assumptions listed above. The results are listed in Table 5.1. For the five-resonator system in the third column of Table 5.1, for example, the load is connected to Resonator-2 first and then to the Resonator-3. Because of the symmetry of the evenly spaced circular systems (see Fig. 5.2), the five-resonator system loaded at Resonator-2 (or Resonator-3) is the same as that loaded at Resonator-5 (or Resonator-4).

Close examination on the computed maximum efficiencies of this range of domino-resonator systems leads to the following observations.

1. When $n$ is even, there exist very low maximum efficiencies when the load is connected to certain resonators. These figures are highlighted in bold-type in Table 5.1. For example, for the four-resonator system in column 2, the maximum efficiency is only 0.07% when the load is connected to Resonator-2 or Resonator-4. For the eight-resonator system in column 6, the maximum efficiencies are close to zero when the load is connected to Resonator-2, -4, -6 or -8.
2. When $n$ is odd, the maximum efficiencies fall within a reasonably close range in all loaded resonators and there does not exist near-zero efficiency in the system.

**Table 5.1** Maximum efficiency of the resonator systems under different loaded positions

| Maximum Efficiency (%)<br>(the sequence numbers of the loaded resonators) | | | | | | | |
|---|---|---|---|---|---|---|---|
| Three-resonator | Four-resonator | Five-resonator | Six-resonator | Seven-resonator | Eight-resonator | Nine-resonator | Ten-resonator |
| 70.68<br>(2, 3) | **0.07**<br>**(2, 4)** | 55.40<br>(2, 5) | 72.78<br>(2, 6) | 44.73<br>(2, 7) | **0.61**<br>**(2, 8)** | 37.79<br>(2, 9) | 59.40<br>(2, 10) |
| | 80.82<br>(3) | 63.61<br>(3, 4) | **2.92**<br>**(3, 5)** | 58.55<br>(3, 6) | 69.20<br>(3, 7) | 54.71<br>(3, 8) | **10.90**<br>**(3, 9)** |
| | | | 72.65<br>(4) | 49.10<br>(4, 5) | **0.07**<br>**(4, 6)** | 39.64<br>(4, 7) | 58.89<br>(4, 8) |
| | | | | | 65.31<br>(5) | 44.39<br>(5, 6) | **1.00**<br>**(5, 7)** |
| | | | | | | | 58.72<br>(6) |

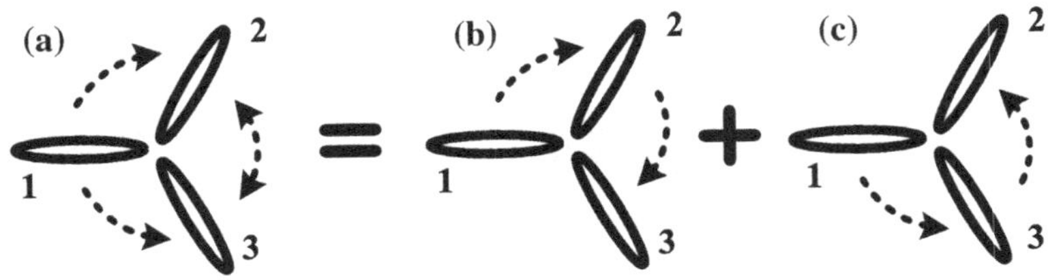


**Fig. 5.5** Two power paths in the circular domino-resonator system

3. Comparison of the systems with odd and even $n$ shows that, while the systems with even number of resonators have some loaded conditions with very low maximum efficiency, they do offer higher maximum efficiency than the odd-numbered counterpart if the appropriate resonators are loaded. For example, for the six-resonator system loaded at Resonator-2 (or -4, -6), the maximum efficiencies are higher than 72%, while the maximum efficiency the three-resonator system only 70.7% noting that the average length of the power paths in the three-resonator system is much shorter.

(iii) **Physical Explanation based on Superposition Analysis and Phasor Diagrams**

The observations made previously can be analyzed with the lumped circuit model and the superposition principle. One interesting feature of the circular domino-resonator systems is that Resonator-1 excited by the AC power source will transfer power in both the clockwise and anticlockwise directions. Starting with the three-resonator system as an example, Fig. 5.5 shows the two power flow paths, with one being clockwise (Fig. 5.5b) and the other anticlockwise (Fig. 5.5c). For a given input AC current $\mathbf{I}_1$, the current phasors in the other resonators in the three-resonator system can be determined with the following equations:

$$j\omega M_{12}\mathbf{I_1} + R_2\mathbf{I_2} + j\omega M_{23}\mathbf{I_3} = 0 \tag{5.1}$$

$$j\omega M_{13}\mathbf{I_1} + j\omega M_{23}\mathbf{I_2} + R_3\mathbf{I_3} = 0. \tag{5.2}$$

By applying the superposition principle, the coupled circuit equations for the clockwise Path-1 in Fig. 5.5b when $M_{13} = 0$ become

$$j\omega M_{12}\mathbf{I_1} + R_2\mathbf{I_{2\_P1}} + j\omega M_{23}\mathbf{I_{3\_P1}} = 0 \tag{5.3}$$

$$j\omega M_{23}\mathbf{I_{2\_P1}} + R_3\mathbf{I_{3\_P1}} = 0. \tag{5.4}$$

Similarly, the corresponding equations for the anticlockwise Path-2 in Fig. 5.5c when $M_{12} = 0$ are

$$R_2\mathbf{I_{2\_P2}} + j\omega M_{23}\mathbf{I_{3\_P2}} = 0 \tag{5.5}$$

$$j\omega M_{13}\mathbf{I}_1 + j\omega M_{23}\mathbf{I}_{2\_P2} + R_3\mathbf{I}_{3\_P2} = 0, \tag{5.6}$$

where $\mathbf{I}_{2\_P1}$ and $\mathbf{I}_{3\_P1}$ represent the currents induced in Resonator-2 and Resonator-3 in Path-1 (Fig. 5.5b); $\mathbf{I}_{2\_P2}$ and $\mathbf{I}_{3\_P2}$ represent the currents induced in the corresponding resonators in Path-2 (Fig. 5.5c). The resultant current phasors for $\mathbf{I}_2$ and $\mathbf{I}_3$ in the three-resonator system are therefore

$$\mathbf{I}_2 = \mathbf{I}_{2\_P1} + \mathbf{I}_{2\_P2} \tag{5.7}$$

$$\mathbf{I}_3 = \mathbf{I}_{3\_P1} + \mathbf{I}_{3\_P2}. \tag{5.8}$$

From (5.4), we can get

$$\mathbf{I}_{3\_P1} = \frac{-j\omega M_{23}}{R_3}\mathbf{I}_{2\_P1}, \tag{5.9}$$

which shows that $\mathbf{I}_{3\_P1}$ should lag behind $\mathbf{I}_{2\_P1}$ by 90°. By substituting (5.9) into (5.3),

$$\mathbf{I}_{2\_P1} = \frac{-j\omega M_{12}}{R_2 + \omega^2 M_{23}^2/R_3}\mathbf{I}_1, \tag{5.10}$$

which implies that $\mathbf{I}_{2\_P1}$ lags behind $\mathbf{I}_1$ by 90°. Similarly for Path-2, $\mathbf{I}_{2\_P2}$ lags behind $\mathbf{I}_{3\_P2}$ by 90° and $\mathbf{I}_{3\_P2}$ lags behind $\mathbf{I}_1$ by 90°. The analysis can be easily extended to an $n$-resonator system and the results can be summarized into one simple rule that the induced current lags behind the source current by 90° in each power path.

The results tabulated in Table 5.1 can then be explained with the help of the vector diagrams showing the phase relationships of the currents. Based on the above analysis, the vector diagrams (not to scale) for the resonator systems with $n = 3$–8 are given in Fig. 5.6a–f, respectively. These vector diagrams provide the physical insights into the behavior of these circular domino-resonator systems.

Careful inspections of the vector diagrams in Fig. 5.6 can offer the explanations to the observations obtained from Table 5.1:

1. The phasor diagram of the four-resonator system shows that the two current phasors due to the two paths in Resonator-3, i.e., $\mathbf{I}_{3\_P1}$ and $\mathbf{I}_{3\_P2}$, are in phase and thus fully re-enforcing each other. However, the current phasor pairs $\mathbf{I}_{2\_P1}$ and $\mathbf{I}_{2\_P2}$ are 180° out of phase and thus canceling each other, and so do $\mathbf{I}_{4\_P1}$ and $\mathbf{I}_{4\_P2}$. This explains why the maximum efficiency of the four-resonator system is virtually zero when either Resonator-2 or Resonator-4 is loaded, and the maximum efficiency is high when Resonator-3 is loaded. Similar arguments can apply to other systems with even number of resonators.
2. The phasor diagrams of the systems with odd number of resonators in Fig. 5.6a, c, and e for $n = 3$, 5 and 7, respectively, indicate that that the two current vectors due to the two power flow paths in each resonator are always perpendicular (i.e.,

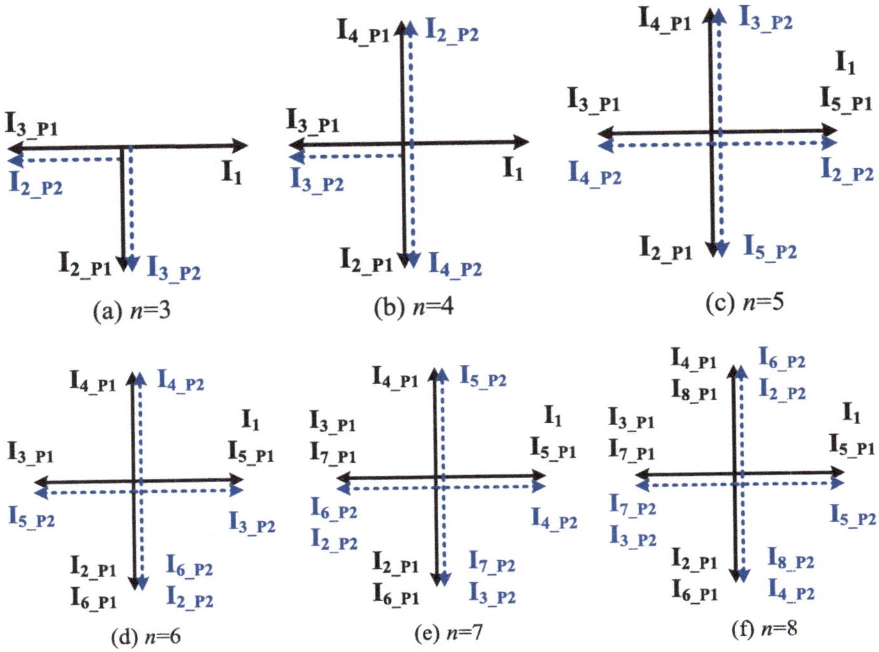


**Fig. 5.6** Vector diagrams of the current vectors for domino-resonator systems with $n = 3$–8 (magnitude not to scale)

$\mathbf{I}_{\mathbf{x\_P1}}$ and $\mathbf{I}_{\mathbf{x\_P2}}$ have a phase shift of 90° for each Resonator-**x**). This means that these two currents in each resonator neither cancel each other nor fully re-enforce each other. This is the reason for the observation that their maximum efficiencies fall fairly close to each other, but they do not have high maximum efficiency as the systems with even number of resonators do.

## 5.4 Optimization of Circular Domino-Resonator Systems with Cross-Couplings

While the simplified analysis in the previous section enables one to understand the interactions of the two power flow paths, it is necessary in many cases to include the cross-coupling effects among the nonadjacent resonators in the analysis in order to determine the optimal operation point for maximum power transfer efficiency. It has been proved in Chap. 4 that the effects of cross-coupling among nonadjacent resonators could shift the optimal operating frequency away from the resonant frequency in a straight domino-resonator system. Besides varying the operating frequency, it is also possible to alter the resonator impedance in order to change the impedance of the power flow paths and optimally direct the power to a targeted resonator.

Two case studies are present below, for a circular three-resonator and four-resonator systems, respectively. The same coils as described in Chap. 4 are used. Three optimizations as defined below are compared.

(1) OPT($R_L$): All the resonators are identical and the system operates at the natural resonant frequency of the resonators, i.e., 519.2 kHz. The load resistance of the system is optimized for maximum efficiency.
(2) OPT($f, R_L$): All the resonators are the same as in (1). The operating frequency and the load resistance are optimized for maximum efficiency. In this optimization, the operating frequency is assumed to change in a small range and thus the coil resistance of the resonators can be considered constant.
(3) OPT($C_i$, $R_L$): All the coils are the same as in (1) and (2) but the compensating capacitances are assumed adjustable. As pointed out in Chap. 3 that the compensation in the source resonator does not affect the efficiency, $C_i$ ($i = 2, 3, \ldots$) represent all the compensating capacitances expect $C_1$. The operating frequency is fixed at 519.2 kHz. The compensating capacitances $C_i$ and the load resistance are optimized for maximum efficiency.

**Cast Study 1: For a Three-Resonator System**

The radius of the circular path $r_{\text{path}}$ is 200 mm which produces a mutual inductance of about 2.647 μH between every two adjacent coils. The efficiency variation with respect to compensating capacitances $C_i$ is plotted in Fig. 5.7. The load resistance is fixed at 12.72 Ω which is the optimal value obtained by doing the optimization OPT($C_i$, $R_L$) with the MATLAB optimization toolbox. A maximum efficiency of 84.04% occurs at the optimized conditions of $C_2 = 0.945C_0$ and $C_3 = 0.995C_0$, where $C_0 = 1.036$ nF is the compensating capacitance of the original resonators. It

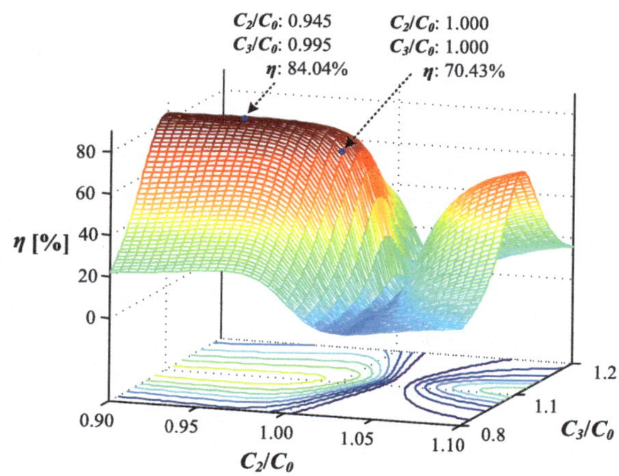


**Fig. 5.7** 3-D plot of the efficiency of the circular three-resonator system as a function as $C_2/C_0$ and $C_3/C_0$

**Table 5.2** Efficiency comparison under different optimization criteria

| | $C_2/C_0$ | $C_3/C_0$ | $f/f_0$ | $R_L$ ($\Omega$) | $\eta$ (%) |
|---|---|---|---|---|---|
| OPT($R_L$) | 1 | 1 | 1 | 14.98 | 70.68 |
| OPT($f$, $R_L$) | 1 | 1 | 0.9855 | 17.05 | 81.00 |
| OPT($C_2$, $C_3$, $R_L$) | 0.945 | 0.995 | 1 | 12.72 | 84.04 |

is noted that the maximum efficiency is much higher than that of the system with original identical resonators (i.e., the efficiency at $C_2/C_0 = C_3/C_0 = 1$).

The predefined three optimizations are done and the results are tabulated in Table 5.2 where $f_0 = 519.2$ kHz. Significant improvements are achieved by optimizing either the operating frequency or the compensating capacitances.

**Case Study 2: For a Four-Resonator System**

The performance of a circular four-resonator system (Fig. 5.8) is also investigated. The mutual inductance between two adjacent resonators of the four-resonator system is also 2.647 $\mu$H with the radius of the circular path $r_{\text{path}}$ of 235 mm. In this case, the cross-coupling mutual inductances between Resonator-1, 3 and 2, 4, (i.e., $M_{13}$ and $M_{24}$) are 0.894 $\mu$H. Table 5.3 and Table 5.4 show the calculated results for the four-resonator system with Resonator-2 and Resonator-3 loaded, respectively.

The results tabulated in Table 5.3 show that the efficiency of the system with Resonator-2 loaded and optimized with OPT($R_L$) is very low due to current cancelation as illustrated in Fig. 5.6. However because the cross-couplings are included in this calculation, the efficiency of OPT($R_L$) is slightly higher than the calculation result in Table 5.1. The results in Table 5.3 indicate that such current cancelation

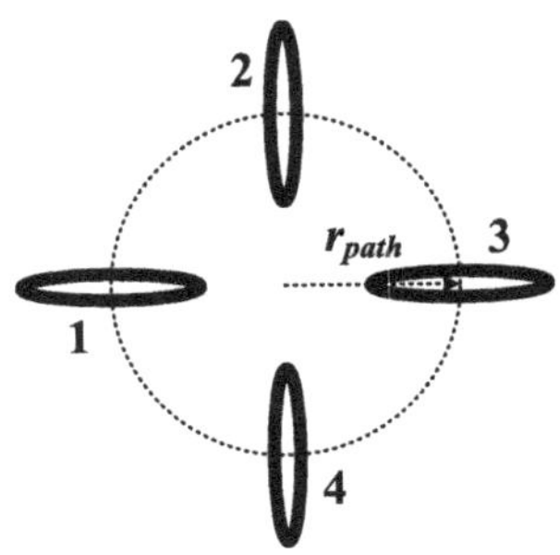


**Fig. 5.8** A circular four-resonator system

**Table 5.3** Performance of the four-resonator system with Resonator-2 loaded under different optimization methods

| | $C_2/C_0$ | $C_3/C_0$ | $C_4/C_0$ | $f/f_0$ | $R_L$ ($\Omega$) | $\eta$ (%) |
|---|---|---|---|---|---|---|
| OPT($R_L$) | 1 | 1 | 1 | 1 | 5.85 | 3.76 |
| OPT($f$, $R_L$) | 1 | 1 | 1 | 0.9736 | 20.56 | 75.00 |
| OPT($C_2$, $C_3$, $C_4$, $R_L$) | 0.9924 | 0.9137 | 0.9137 | 1 | 11.57 | 83.01 |

**Table 5.4** Performance of the four-resonator system with Resonator-3 loaded under different optimization methods

| | $C_2/C_0$ | $C_3/C_0$ | $C_4/C_0$ | $f/f_0$ | $R_L$ (Ω) | $\eta$ (%) |
|---|---|---|---|---|---|---|
| OPT($R_L$) | 1 | 1 | 1 | 1 | 16.60 | 78.68 |
| OPT($f$, $R_L$) | 1 | 1 | 1 | 0.9885 | 18.07 | 80.68 |
| OPT($C_2$, $C_3$, $C_4$, $R_L$) | 0.9712 | 0.9920 | 0.9712 | 1 | 16.56 | 81.73 |

can be eliminated either by deviating the operating frequency from the resonant frequency or by altering the impedance of the resonators (via changes in capacitor values). The theoretical results show that an efficiency of 75% and 83% can be achieved by doing optimizations of OPT($f$, $R_L$) and OPT($C_i$, $R_L$), respectively. When the load is connected to Resonator-3, the theoretical results are included in Table 5.4. The efficiency improvements are not significant as in the previous case since there is no current cancelation in this loading position and the cross-coupling effect is not dominant.

## 5.5 Practical Verification

Experiments have been carried out with a circular four-resonator system (Fig. 5.9). The radius of the circular paths ($\boldsymbol{r}_{\mathbf{path}}$) of the four-resonator systems is 235 mm. Resonator-1 is driven by a RF Power Amplifier with magnitude and frequency controlled. The currents in each resonator are recorded by a digital storage oscilloscope so that their phasor diagrams can be produced.

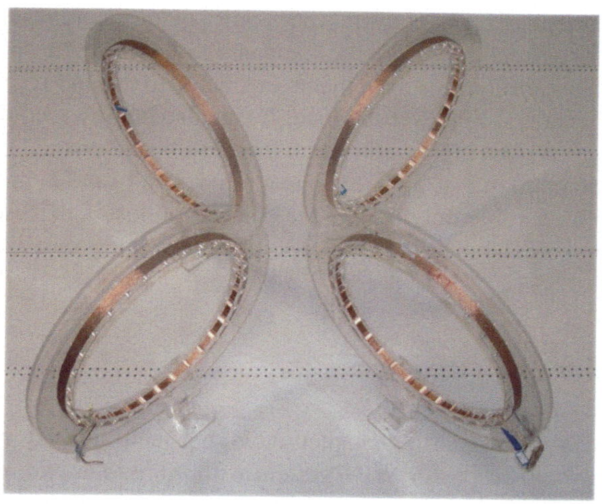

**Fig. 5.9** A circular four-resonator system

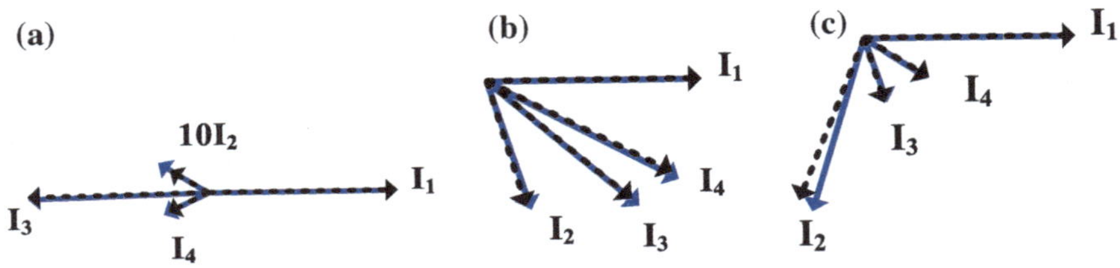


**Fig. 5.10** Current phasors of the circular four-resonator system with Resonator-2 loaded and the current in Resonator-1 as vector **1**

Experiments are conducted to evaluate the system with Resonator-2 loaded, under three different sets of operation conditions.

(a) Identical resonators with a natural resonant frequency of 519.2 kHz. The load resistance $R_L = 21.20\ \Omega$. The operating frequency $f = 520$ kHz which is close to the theoretical $f_0$ of 519.2 kHz.
(b) OPT($f$, $R_L$): The operating frequency $f = 0.9736 f_0 = 505$ kHz which is the optimal value. $R_L = 21.20\ \Omega$ which close to the theoretical optimal value of 20.56 Ω.
(c) OPT($C_i$, $R_L$): The operating frequency $f = 520$ kHz. $C_2 = C_0$ which is close to the theoretical value of $0.9924C_0$. $C_3 = C_4 = 0.913C_0$ which is close to the theoretical optimal value of $0.9137C_0$. $R_L = 11.82\ \Omega$ which is close to the theoretical optimal value of 11.57 Ω.

The current phasors of the resonators under the above three different test conditions are shown in Fig. 5.10a, b, and c, respectively, with $\mathbf{I}_1$ normalized as **1**. It should be noted that the relative magnitude of $\mathbf{I}_2$ is very small in test (a), therefore, $10\mathbf{I}_2$ with 10 times the magnitude is shown in Fig. 5.10a for easy inspection. The solid vectors are measurements and the dotted ones are theoretical predictions. The measured result confirms the validity of the theoretical analysis. By shifting the operating frequency from the resonant frequency to the optimized frequency, the phasor diagram changes from that in Fig. 5.10a to that in Fig. 5.10b. A much larger $\mathbf{I}_2$ is now generated to power the load. It can also be observed that optimizing the capacitors offers even higher efficiency. This can be understood from the corresponding vector diagram in Fig. 5.10c. The magnitudes of $\mathbf{I}_3$ and $\mathbf{I}_4$ are smaller in Fig. 5.10c than in Fig. 5.10b, implying that the conduction losses in Resonator-3 and Resonator-4 are lower. The measured and theoretical efficiency curves are shown in Fig. 5.11.

## 5.6 Discussion

Chapter 4 and this chapter show that there will be more than one power paths in a multi-resonator WPT system with more than two resonators. These power paths might impose canceling electromagnetic forces on the loaded coil if the resonators have an identical resonant frequency and the operating frequency equals to this

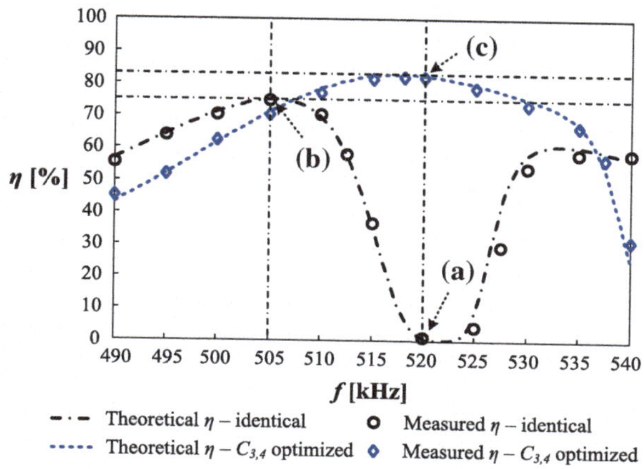


**Fig. 5.11** Efficiency curves of the circular four-resonator systems with Resonator-2 loaded under the test conditions (a), (b), and (c)

resonant frequency as what happens in a two-resonator system. This might lead to very low power transfer efficiency. Therefore, the power flows in a multi-resonator system should be optimized either by shifting the operating frequency away from the resonant frequency or adjusting the individual resonant frequencies of the resonators by changing their compensating capacitances. Apparently, the latter method can provide better performance since it offers more degrees of freedom for optimization.

## Reference

1. Zhong WX, Lee CK, Hui SY (2012) Wireless power domino-resonator systems with non-coaxial axes and circular structures. IEEE Trans Power Electron 27(11):4750–4762

# Chapter 6
# A Method to Create More Degrees of Freedom for Designing WPT Systems—Coil Splitting

## 6.1 Introduction

It has been proved in Chap. 4 that adding relay resonators in between the transmitter and receiver can substantially increase WPT efficiency. But in many real applications, it is not practical to insert a relay resonator in the middle of the power transfer path. The alternative way is to place the relay resonator on the same side of the transmitting coil or the receiving coil.

This chapter presents a study on a WPT system with its transmitting coil split into two, i.e., forming a new transmitting coil and the relay coil. The splitting and compensation schemes are analyzed in detail. It is found that after properly splitting the transmitting coil and optimally compensating the resultant coils, the efficiency can be improved compared to that before splitting. This is because the input impedance seen by the power source is changed by using this coil-splitting method. In Chap. 10, it is demonstrated that this coil-splitting method could be used to alter the optimal load resistance of a WPT system. Generally speaking, this coil-splitting method provides additional freedom for designing and optimizing a WPT system.

More studies on the similar concept of using intermediate coil(s) can be found in [1, 2].

## 6.2 Theoretical Analysis

As shown in Fig. 6.1, Resonator-$a$ is added to the transmitter side of a two-resonator system denoted with 1 and 2. Now Coil-1 becomes a relay. For simplicity, the magnetic coupling between Coil-$a$ and the receiver Coil-2 is assumed to be negligibly small when compared with the coupling between Coil-1 and Coil-2. This condition is valid when Coil-$a$ has much less turns compared with Coil-1. If the resonators are excited at the resonant frequency making $X_a = X_1 = X_2 = 0$, the lumped circuit equations of the system are



W. Zhong et al., *Wireless Power Transfer*, CPSS Power Electronics Series,
https://doi.org/10.1007/978-981-15-2441-7_6

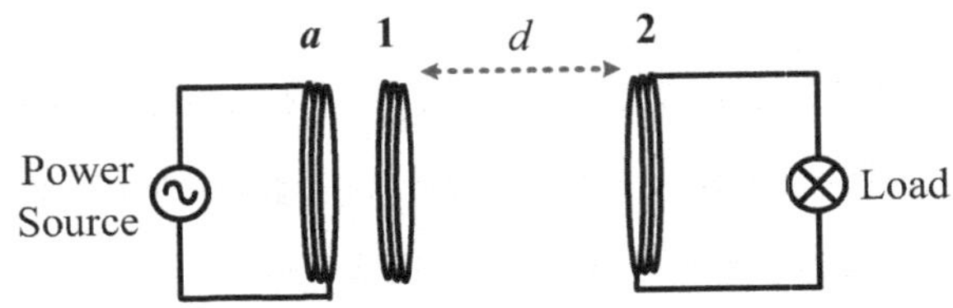


**Fig. 6.1** A WPT system with an additional resonator at the transmitter side

$$(R_S + R_{Pa})\mathbf{I}_a + j\omega M_{a1}\mathbf{I}_1 = \mathbf{V}_S \tag{6.1}$$

$$j\omega M_{a1}\mathbf{I}_a + R_{P1}\mathbf{I}_1 + j\omega M_{12}\mathbf{I}_2 = 0 \tag{6.2}$$

$$j\omega M_{12}\mathbf{I}_1 + R_2\mathbf{I}_2 = 0, \tag{6.3}$$

where $R_{Pa}$ is the parasitic resistance of the new primary Resonator-$a$; $M_{a1}$ is the mutual inductance between Coil-$a$ and Coil-1.

The overall energy efficiency of the three-resonator system is

$$\eta' = \eta_a \eta_1' \eta_2, \tag{6.4}$$

where $\eta_a = \frac{R_{Ea}}{R_S + R_{Pa} + R_{Ea}}$ is the efficiency of Resonator-$a$ while $R_{Ea} = \frac{\omega^2 M_{a1}^2}{R_{P1} + R_{E1}}$ is the reflected resistance from Coil-1 to Coil-$a$; $R_{E1} = \frac{\omega^2 M_{12}^2}{R_2}$ is the reflected resistance from the receiver to Coil-1; $\eta_1' = \frac{R_{E1}}{R_{P1} + R_{E1}}$ and $\eta_2 = \frac{R_L}{R_L + R_{P2}}$ are the efficiencies of Coil-1 and Coil-2, respectively.

For the original two-resonator system with Coil-1 and Coil-2 (i.e., before adding Coil-$a$), the efficiency is given by

$$\eta = \eta_1 \eta_2, \tag{6.5}$$

where

$$\eta_1 = \frac{R_{E1}}{R_S + R_{P1} + R_{E1}} \tag{6.6}$$

$$\eta_2 = \frac{R_L}{R_L + R_{P2}}. \tag{6.7}$$

To ensure the efficiency of the three-resonator system higher than that of the original two-resonator system, the following inequality should be met:

$$\eta' > \eta, \tag{6.8}$$

which is equivalent to

$$|M_{a1}| > \frac{R_{P1} + R_{E1}}{\omega}\sqrt{1 + \frac{R_{Pa}}{R_S}}. \tag{6.9}$$

The inequality (6.9) specifies the required conditions for the efficiency of the three-resonator system to be higher than that of the original two-resonator system before adding Coil-$a$. It implies that as long as $M_{a1}$ is large enough to meet the inequality of (6.9), a three-resonator system can enjoy higher efficiency than a two-resonator system for the same power transfer distance.

## 6.3 Computer-Aided Analysis and Verifications

The specified operating conditions defined by Eq. (6.9) can be implemented in a three-resonator system. As shown in Fig. 6.2, Coil-$a$ and Coil-1 are obtained by splitting the transmitting coil of a conventional two-resonator system. It should be noted that Coil-$a$ or Coil-1 could be either the inner coil or the outer coil. When Coil-$a$ and Coil-1 are connected in series, the three-resonator system will become the original two-resonator system and thus, it will be an appropriate comparison between these two systems whose only difference is the connection method.

In this example, the transfer distance is set at 30 mm. $W_r$ represents the coil width of the receiving coil as shown in Fig. 6.3. This coil width is defined as the difference

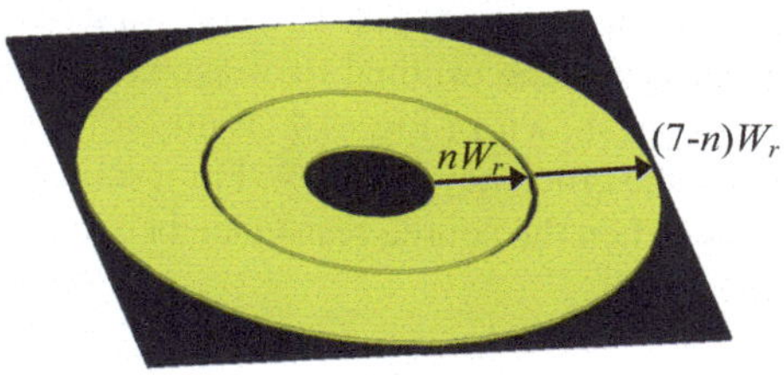


**Fig. 6.2** The three-coil structure

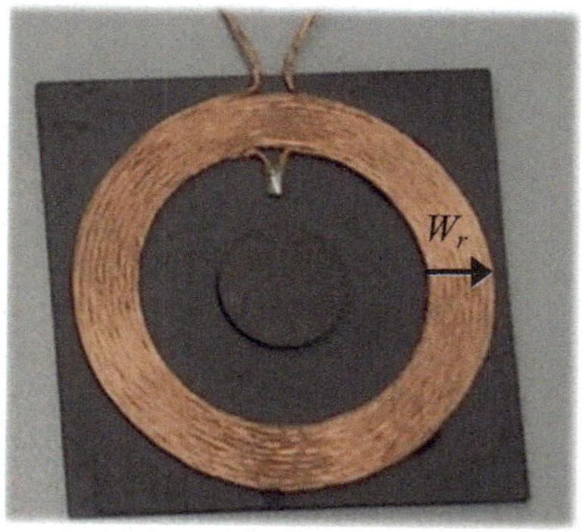


**Fig. 6.3** The receiver of the practical WPT system

**Table 6.1** Parameters of a practical receiving coil

| Parameters | Symbol | Value |
|---|---|---|
| Inner diameter | $d_i$ | 21.7 mm |
| Coil width | $W_r$ | 5.32 mm |
| Outer diameter | $d_o$ | 27.02 mm |
| Number of turns per layer | – | 9 |
| Number of layers | – | 2 |

of the outer radius of the coil and the inner radius of the hollow. The transmitting coil is assumed to have the same inner hollow radius as the receiving coil does. The simulations are based on the parameters of the practically receiving coil in Fig. 6.3 and the parameters are shown in Table 6.1. A litz wire with 24 strands and a strand diameter of 0.08 mm is used.

In the example of Fig. 6.2, the driving and the relay coils are located at the transmitter side. One of the coils is connected to the power source while the other one operates as a relay. In order to determine the optimal arrangements of these two coils, the original transmitting coil with a coil width of $7W_r$ is divided into two coils. The coil widths of the inner coil and the outer coil are $nW_r$ and $(7 - n)W_r$, respectively. If the inner coil is used as the driving coil, then the outer coil will be used as the relay coil, and vice versa. For the relay coil, a capacitor is connected in series with the coil to form a coil-resonator (i.e., LC resonator).

The inductance and the AC resistance values are obtained with the help of a Finite-Element-Analysis (FEA) software and the method used to calculate AC resistance of a coil is introduced in [3]. The operating frequency is set at 115.6 kHz. The calculated efficiency of the original two-resonator system (i.e., coil-*a* and coil-1 connected in series) is shown in Fig. 6.4. With an ideal power source (i.e., source resistance is zero), the efficiency is 75.3%. For a source impedance of 2 Ω, the efficiency is about 70%. When the source resistance becomes 50 Ω, the efficiency is 24%.

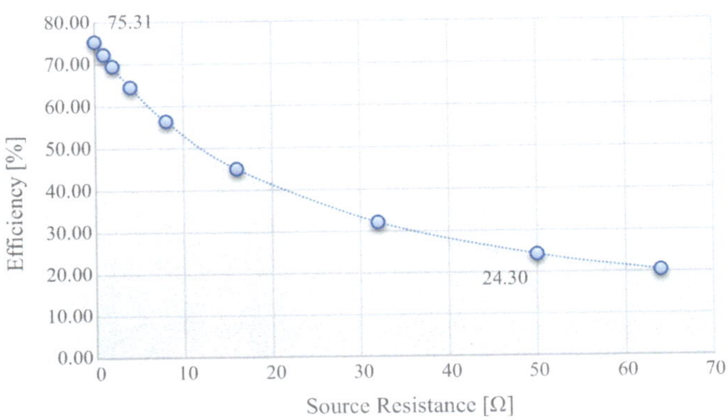


**Fig. 6.4** Efficiency of the original two-resonator system as a function of the source resistance

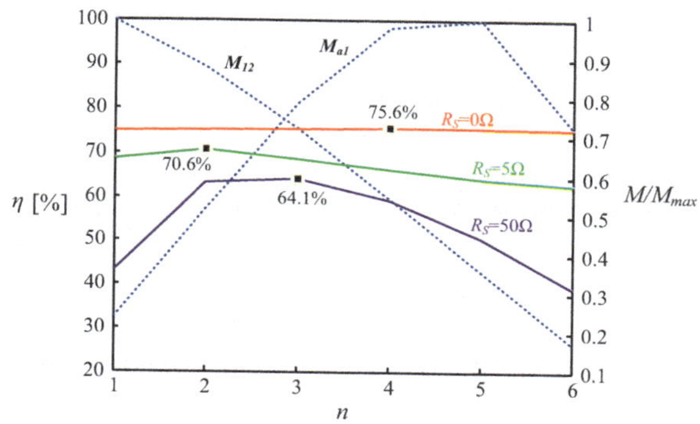


**Fig. 6.5** Mutual inductance ($M_{a1}$ and $M_{12}$) curves and efficiency curves (for source resistance $R_s = 0$, 5, and 50 Ω) with different sizes for the inner and outer coils when the inner coil is excited as the driving coil

### 6.3.1 Use of the Inner Coil as Coil-a and the Outer Coil as Coil-1

The first set of simulations has been conducted with the inner coil connected to the power source (as Coil-*a*) and the outer coil connected in series with a capacitor to form a relay resonator (Coil-1). For the coil widths of the inner coil and the outer coil as $nW_r$ and $(7 - n)W_r$, respectively, the maximum efficiencies of the three-resonator system as a function of $n$ under a range of source resistance are shown in Fig. 6.5.

When the source resistance is assumed zero, the three-resonator system can achieve basically the same efficiency as the original two-resonator system does, no matter how the transmitter coil is cut. When the source resistance becomes larger, the importance of a large $M_{a1}$ becomes more obvious as predicted by (6.9). With a larger $M_{a1}$, a smaller current is required in Coil-*a* in order to generate a large enough current in Coil-1 for power transfer to Coil-2. However, a large magnetic coupling between the relay resonator and the receiving coil ($M_{12}$) also helps to achieve a high efficiency and this is the reason why the efficiency curve of the system with 50 Ω source resistance reaches its peak value at a point which has a relatively large $M_{a1}$ and $M_{12}$ at the same time. On the other hand, when the source resistance is negligibly small, the effect of $M_{a1}$ is smaller. $M_{12}$ and $M_{a1}$ in the figure are normalized to their respective maximum values.

### 6.3.2 Use of the Inner Coil as Coil-1 and Outer Coil as Coil-a

In the second set of simulations, the outer coil is connected to the power source and thereby, the outer coil is Coil-*a* and the inner coil is Coil-1. Now the efficiency of the three-resonator system with a 50 Ω source resistance is higher than that in the first case, as shown in Fig. 6.6. The reason is that $M_{a1}$ and $M_{12}$ in this case increase simultaneously which is different compared to that in the first case as shown in

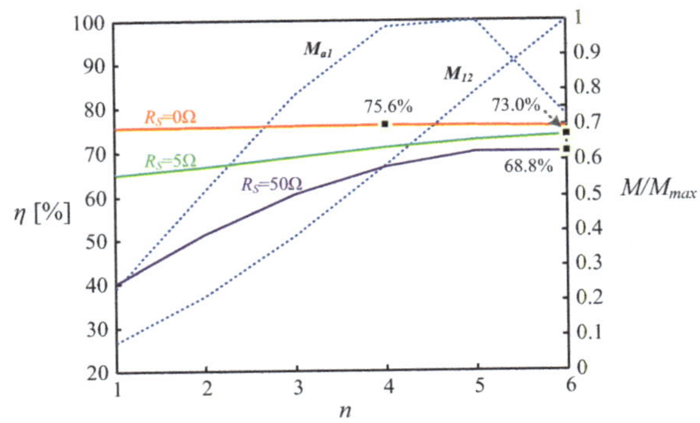


**Fig. 6.6** Mutual inductance ($M_{a1}$ and $M_{12}$) curves and efficiency curves (for source resistance $R_s = 0$, 5, and 50 Ω) with different sizes for the inner and outer coils when the outer coil is excited as the driving coil

Fig. 6.5. Both $M_{a1}$ and $M_{12}$ can reach higher values when $n$ is large and thus a higher efficiency can be achieved.

### *6.3.3 Choice of Structures and Effects of Source Impedance*

By comparing the results of these two groups of simulations, the optimal $n$ which achieves maximum efficiencies with a 5 Ω or 50 Ω source resistance is 6. In other words, the inner coil has 6 times $W_r$ while the outer coil has 1 $W_r$ (see Fig. 6.10). Then a comparison is made under different loading conditions between the three-resonator system with this optimal structure and the original two-resonator system which has a $7W_r$ transmitting coil.

In a real implementation, the input power source feeding the transmitting coil is likely consisting of, from AC mains, an input filter, a rectifier, a PFC converter, and a high-frequency inverter. So the source resistance might range from 0 to 5 Ω. Theoretical comparisons on the cases of a source impedance of 0, 5, and 50 Ω have been studied for a two-resonator and a three-resonator system. The results are recorded in Figs. 6.7, 6.8, and 6.9. When the source impedance is 0 Ω, there is basically little performance difference between the two-resonator and three-resonator systems. While Figs. 6.8 and 6.9 show that the efficiency of the two-resonator system will deteriorate much faster than the three-resonator one when the source impedance increases.

## 6.4 Experimental Verification

Experiments have been carried out using a practical three-coil system as shown in Fig. 6.10. The practical parameters of Coil-$a$ and Coil-1 are listed in Table 6.2. The electrical parameters of the coils are calculated by using the Maxwell Finite-element software and are listed in Table 6.3.

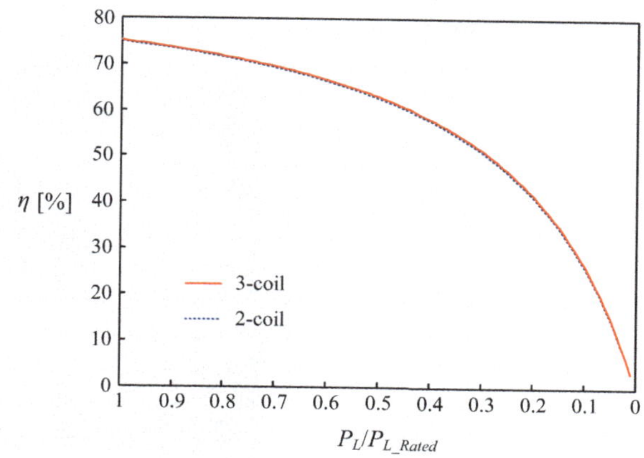


**Fig. 6.7** Efficiency comparison between the three-resonator and the two-resonator systems under different loading conditions when source resistance is 0 Ω

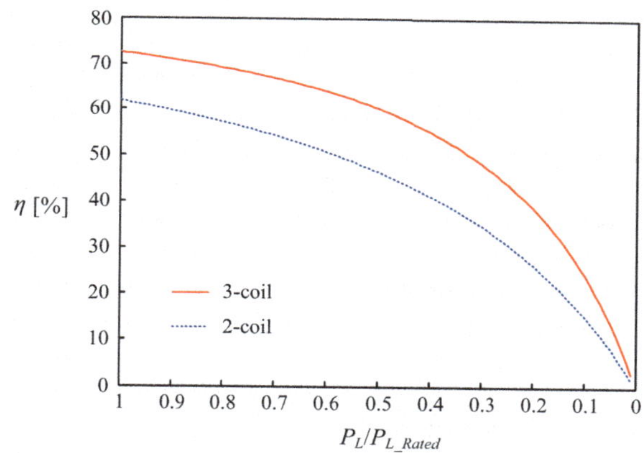


**Fig. 6.8** Efficiency comparison between the three-resonator and the two-resonator systems under different loading conditions when source resistance is 5 Ω

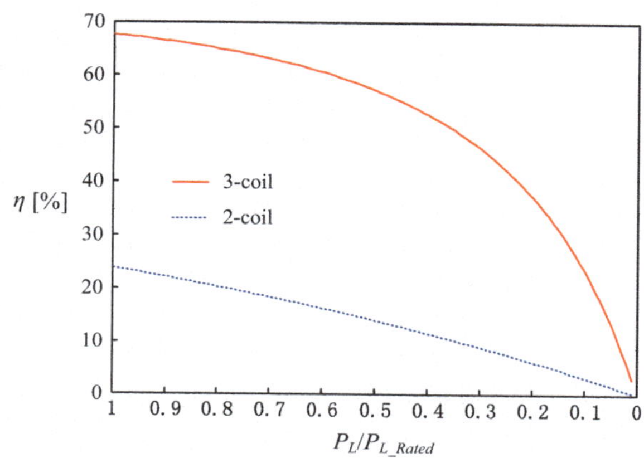


**Fig. 6.9** Efficiency comparison between the three-resonator and the two-resonator systems under different loading conditions when source resistance is 50 Ω

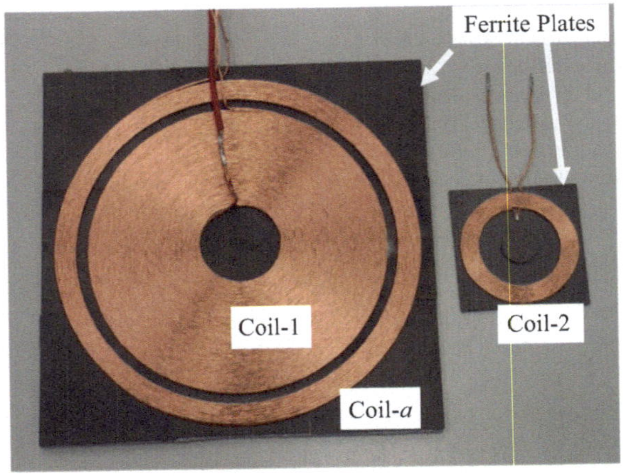


**Fig. 6.10** The three-coil system used in experiments

**Table 6.2** Parameters of the practical coils on the transmitter side

| Parameters | Coil-*a* (power coil) | Coil-1 (repeater) |
|---|---|---|
| Inner diameter | 21.7 mm + 6 × $W_r$ = 85.54 mm | 21.7 mm |
| Axial width | 1 × $W_r$ = 5.32 mm | 29.9 mm (designed value: 6 × $W_r$ = 31.92 mm) |
| Number of turns per layer | 9 | 6 × 9 = 54 |
| Number of layers | 2 | 2 |

**Table 6.3** Parameters of the coils obtained from FEA

| | Values |
|---|---|
| $L_a$ | 114.1 μH |
| $R_a$ | 1.05 Ω |
| $L_1$ | 1.050 mH |
| $R_1$ | 4.1 Ω |
| $L_2$ | 23.13 μH |
| $R_2$ | 0.30 Ω |
| $M_{a1}$ | 142.8 μH |
| $M_{12}$ | 14.73 μH |
| $M_{a2}$ | 2.866 μH |

The compensating capacitance of the receiving coil (Coil-2) is set at about 82 nF which is also close to the optimal value for an operating frequency of 115.6 kHz in this study. The optimal compensating capacitance of Coil-1 is 1.70 nF and 1.83 nF for 0 Ω and 50 Ω source resistance, respectively. The efficiencies of the three-resonator system are compared with those of the two-resonator system (in which

Coil-*a* and Coil-1 in the three-resonator system are connected in series to form a single transmitting coil).

### 6.4.1 Efficiency Evaluation

The comparisons are provided in Fig. 6.11 and Fig. 6.12 for a source resistance of 2 Ω and 50 Ω, respectively. In the simulations, the core losses and the capacitor losses are neglected and therefore, the differences between the measured results and the simulation results are obvious but the trends are similar. As expected, the three-resonator system offers a higher efficiency, especially when the source resistance is significant.

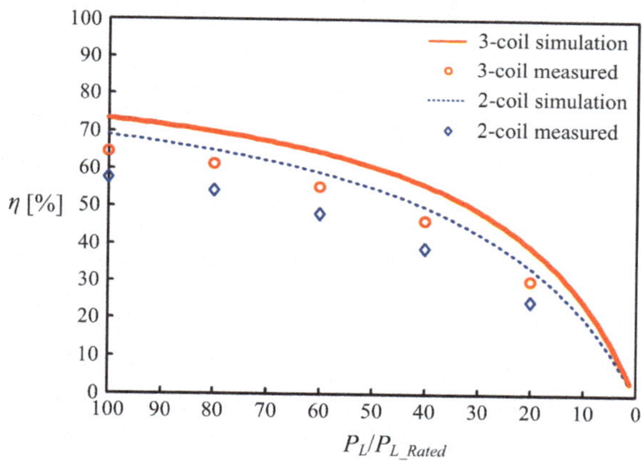


**Fig. 6.11** Measured and simulated efficiencies of the three-resonator and two-resonator systems under different loading conditions when the source resistance is 2 Ω

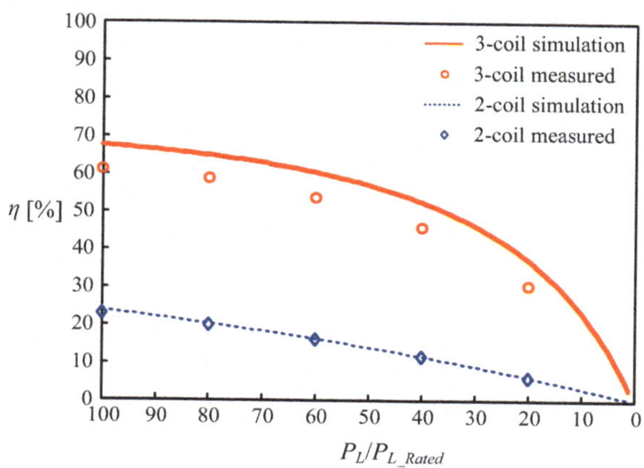


**Fig. 6.12** Measured and simulated efficiencies of the three-resonator and two-resonator systems under different loading conditions when the source resistance is 50 Ω

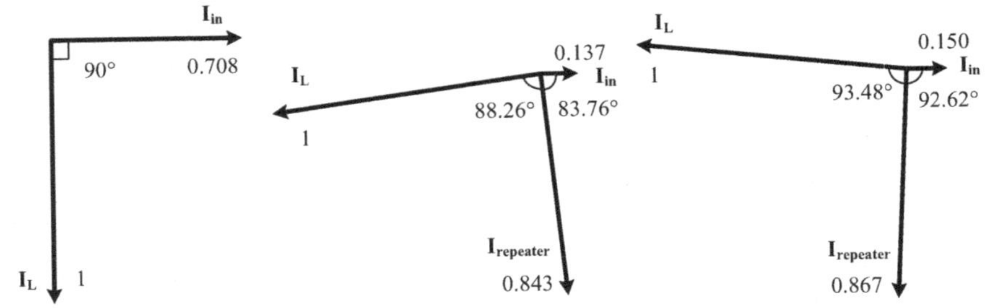


(a) two-coil simulated (b) three-coil simulated (c) three-coil measured

**Fig. 6.13** Per-unit current vector diagrams of **a** the two-resonator system, **b** and **c** three-resonator system; with a source resistance of 50 Ω for both systems

## *6.4.2 Current Stress Evaluation*

An effective way to highlight the advantage of the proposed theory is to check the current stress of the driving circuits in the transmitting coils of the two-resonator and three-resonator systems for the same load currents in the receiving coils. The per-unit vector diagrams are used here for comparison purposes. With the load current taken as 1.0 per unit, Fig. 6.13 shows the per-unit vector diagrams of the two-resonator system and the three-resonator system, for an input source impedance of 50 Ω. The driving current in the two-resonator system is 0.708 per unit in simulations and 0.698 in measurements, while that of the three-resonator system is only 0.15 from measurement results. The relay coil current $I_2$ in the three-resonator system is 0.867, indicating that a large magnetic flux is generated by the relay coil current for power transfer. Therefore, the features of both current stress reduction and improved efficiency have been practically confirmed.

## 6.5 Conclusion

The focus of this chapter is to describe a methodology to provide another degree of freedom for designing a WPT system. By controlling the division ratio of the primary winding, the impedance of the WPT system can be properly tuned and the efficiency of the resulting three-resonator system can be more efficient than a two-resonator counterpart. The theory and design criteria for achieving this objective have been presented and explained. The key features of the three-resonator system design are to adjust the impedance of the WPT system and shift the current stress from the driving circuit and coil to the relay coil or vice versa.

## References

1. Moon S, Kim B, Cho S, Ahn C, Moon G (2014) Analysis and design of a wireless power transfer system with an intermediate coil for high efficiency. IEEE Trans Ind Electron 61(11):5861–5870
2. Mirbozorgi SA, Yeon P, Ghovanloo M (2017) Robust wireless power transmission to mm-sized free-floating distributed implants. IEEE Trans Biomed Circuits Syst 11(3):692–702
3. Sullivan CR (2001) Computationally efficient winding loss calculation with multiple windings, arbitrary waveforms, and two-dimensional or three-dimensional field geometry. IEEE Trans Power Electron 16(1):142–150

# Part III
# Maximum Efficiency Operation

Magnetic resonance WPT systems can achieve maximum efficiency only at their optimal load resistance. For the most common SS WPT systems, such optimal load resistance occurs at a fixed value that depends on the mutual inductance and the resonator resistances only. However, the mutual inductance and the load resistance may vary in many WPT applications, making it difficult to keep the equivalent load resistance fixed at the optimal value for achieving maximum-efficiency power transfer. This part focuses on the techniques that are able to improve WPT efficiency by matching the actual equivalent load resistance to the optimal value of a WPT system.

# Chapter 7
# Review of Maximum-Efficiency-Operation Techniques

In the design stage of a WPT system, many aspects including maximum achievable efficiency, misalignment tolerance, stray magnetic fields, etc., should be considered. However, after a WPT system is designed and fabricated, its maximum achievable efficiency is determined. Yet the system might not be able to operate at this maximum efficiency because the actual efficiency of a WPT system is highly dependent on the mutual coupling and the equivalent load resistance which are varying [1, 2].

To ensure a WPT system to operate at the maximum achievable efficiency regardless of the mutual coupling and the load resistance, the actual load resistance should be matched to the optimal load resistance of the system in real time. This chapter reviews some technical approaches for achieving maximum or near-maximum efficiency operations in WPT systems. These methods are classified in terms of their operating principles, circuitry, and control. Their advantages and disadvantages are compared.

## 7.1 Theory of Maximum-Efficiency-Operation WPT

One of the inherent characteristics of magnetic resonance WPT systems is that the maximum-efficiency-operation point occurs only when the load resistance equals to a specific optimal value which depends on the parameters of the system. When the load resistance shifts away from this optimal value, the system efficiency becomes lower than the highest possible value. Take the commonly used series–series (SS) compensated two-resonator WPT system as an example. As provided in Chap. 2, the optimal load resistance and the maximum efficiency are given by

$$R_{L\_OPT} = R_2\sqrt{1 + \frac{\omega^2 M_{12}^2}{R_1 R_2}} = R_2\sqrt{1 + k_{12}^2 Q_1 Q_2} \tag{7.1}$$


W. Zhong et al., *Wireless Power Transfer*, CPSS Power Electronics Series,
https://doi.org/10.1007/978-981-15-2441-7_7

$$\eta_{\max} = \frac{\frac{\omega^2 M_{12}^2}{R_1 R_2}}{\left(1 + \sqrt{1 + \frac{\omega^2 M_{12}^2}{R_1 R_2}}\right)^2} = \frac{k_{12}^2 Q_1 Q_2}{\left(1 + \sqrt{1 + k_{12}^2 Q_1 Q_2}\right)^2}, \tag{7.2}$$

where $k_{12} = \frac{M_{12}}{\sqrt{L_1 L_2}}$ is the coupling coefficient between two windings; $Q_i = \frac{\omega L_i}{R_i}$ ($i$ = 1 or 2) is the quality factor of the Resonator-$i$.

From (7.1), we know that this optimal load resistance depends on the resistances of the resonators, the mutual inductance, and also the operating frequency. This unique feature of WPT systems raises two critical issues. Firstly, when the equivalent load resistance ($R_L$) at rated power is not equal to the optimal value ($R_{L\_OPT}$), the system operates at a lower efficiency at rated power. Secondly, in many practical applications such as charging a battery load, the load resistance may vary with time. Even if we could set the optimal load resistance to the equivalent load resistance at rated power when designing the system, the actual equivalent load resistance is time-varying and will shift away from the optimal value. Therefore, for those applications where load resistance or the parameters of the system will vary, the average efficiency could be much lower than the highest possible efficiency. Hence, maximum-efficiency operation (MEO) techniques have been proposed to enable WPT systems to operate at the maximum or near-maximum efficiency regardless of the load resistance or parameter variations.

## 7.2 Factors Affecting Maximum-Efficiency-Operation

### 7.2.1 *Efficiency Degradation Due to Variations in Magnetic Coupling*

For wireless EV charging, there might be variations in the horizontal coil misalignment and also the air gap between the transmitting and receiving coils. Thus the coupling coefficient will change substantially [3]. According to (7.1), the changing mutual inductance will alter the optimal load resistance. Without impedance matching, the system might operate with low efficiency for some positions. A case study on the stationary EV wireless charging is provided to illustrate the possible efficiency drop due to the variations in magnetic coupling. Assuming that the quality factor of the coils and the resistance of the coils are 300 and 1 Ω, respectively, and the coupling coefficient varies from 0.05 to 0.3, Fig. 7.1 shows the efficiency curves of the system with the maximum and minimum coupling coefficients. If the load resistance at the rated power is set to the optimal value at the minimum coupling, i.e., 15 Ω, the system will operate with an efficiency of 93.57% at the maximum coupling without any MEO technique. However, if MEO can be achieved, the system efficiency should become close to 97.8% which is 4% higher. If the load resistance at the rated power is set to the optimal value at the maximum coupling (i.e., 90 Ω),

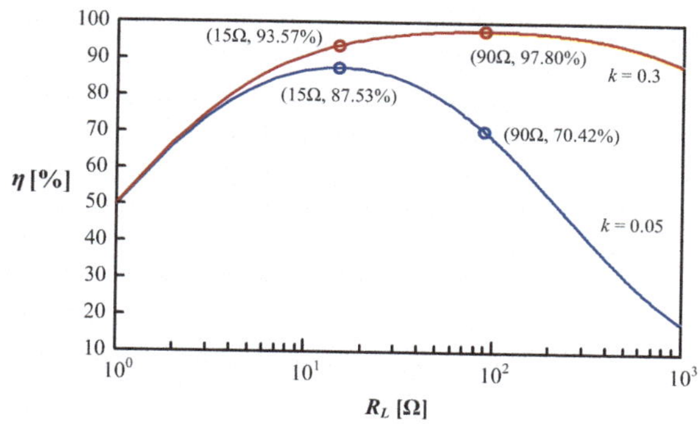


**Fig. 7.1** Efficiency curves under different coupling conditions

the efficiency reduction at the minimum coupling would be as large as 17%. One conclusion that can be drawn from the case study is that the load resistance at the rated power should be set to the optimal value at the minimum coupling, if possible. However, it might be difficult to do so in practice when a specific load is given, and the prior design objective is to maximize the efficiency at worst coupling rather than to set the optimal load resistance equal to the actual load resistance.

## 7.2.2 *Efficiency Degradation Due to Load Resistance Variation*

WPT has found applications in a range of battery loads such as EV, consumer electronics, and implantable electronics. In the battery charging process, the equivalent load resistance will vary and so it is a challenge to maintain MEO. Figure 7.2a shows the typical charging voltage and current profile for Li-ion batteries. The battery needs a constant current (CC) in the first charging stage and constant voltage (CV) in the second stage. In the CC mode, the voltage of the battery will increase slowly. Thus the load resistance only changes moderately, and so the efficiency variation in CC mode

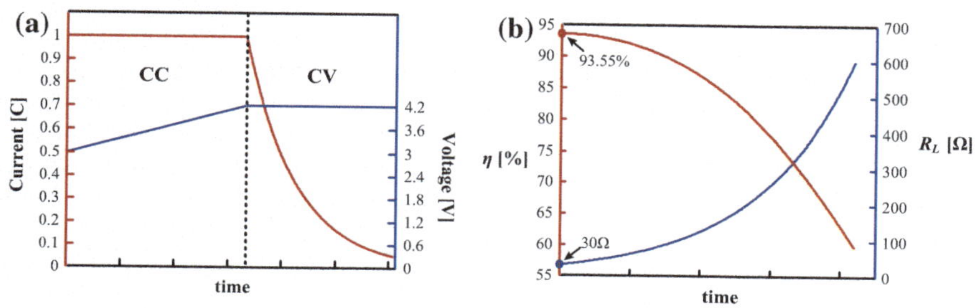


**Fig. 7.2** **a** Typical charging profile of Li-ion batteries; **b** load resistance and efficiency variation in the CV mode

is not significant. But in the CV mode, the charging voltage is constant and the charging current will decrease from the rated value to a much smaller value. Therefore, the load resistance and the resulting efficiency in CV mode will vary substantially as shown in Fig. 7.2b. It is assumed that the charging system is optimized at the rated power point (i.e., the transition point between CC and CV) by designing the optimal load resistance of the system equal to the load resistance at the rated power. The efficiency curve showed in Fig. 7.2b is worked out based on a charging system with a coil resistance of 1 Ω, a coil quality factor of 300, and a coupling coefficient of 0.1. By integrating and averaging the efficiency over the whole CV charging period, the average efficiency can be obtained as 83% which is 10.5% less than the maximum efficiency. The efficiency improvement in the whole charging process will depend on the time ratio of the CC and CV charging modes.

## 7.3 Review of MEO Strategies

### *7.3.1 Using Standard DC–DC Converters on the Receiver Side*

Standard DC–DC converters can transform an impedance to another value in DC as transformers can perform impedance transformation in AC. So a straightforward solution for the mismatch between the actual load resistance and the optimal load resistance of a given WPT system is to use a DC–DC converter after the AC–DC rectifier in the receiver side [1, 4–13]. The load resistance can then be transformed to the optimal value at which the system can achieve the highest efficiency. The system configuration is shown in Fig. 7.3.

Although a DC–DC converter is widely adopted in the receiver side by WPT systems, it is normally used to regulate the output voltage, current, or power. One of the early reports on using DC–DC converters for impedance matching was found in [4]. The input impedance of the DC–DC converter, i.e., $R'_L$ in Fig. 7.3, is a function of the duty cycle of the converter and the load resistance.

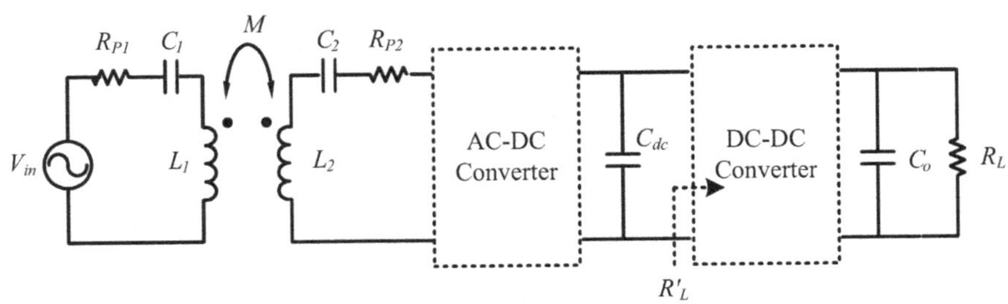


**Fig. 7.3** A WPT system with a DC–DC converter for load resistance transformation

For buck converter ($R'_L \geq R_L$):

$$R'_L = \frac{R_L}{D^2} \tag{7.3}$$

For boost converter ($R'_L \leq R_L$):

$$R'_L = (1 - D)^2 R_L \tag{7.4}$$

For buck–boost converter:

$$R'_L = \left(\frac{1 - D}{D}\right)^2 R_L, \tag{7.5}$$

where $D$ ($0 < D < 1$) is the duty cycle and the converters are assumed to operate in the continuous conduction mode.

By adjusting the duty cycle of the converter, load resistance can be converted to a value either larger (buck) or smaller (boost). The control schemes of the system will be discussed in detail in Sect. 7.4.

### 7.3.2 Using Boost-Type Converters on the Receiver Side

In [14–24], several boost-type converters have been suggested to realize MEO for series-compensated-secondary WPT systems. Some typical topologies to implement boost mechanism are shown in Fig. 7.4. These receiver circuits share the same basic idea that the load is disconnected from the power source periodically by using the active switches. Therefore, the RMS value of the rectifier input voltage is lower than what normally needs when the load is continuously connected (i.e., $\frac{2\sqrt{2}V_O}{\pi}$, where $V_O$ is the DC output voltage). This is essentially a *boost* function and thereby the secondary converters in Fig. 7.4 are termed boost-type converters. As a result, they are all able to convert the actual load resistance to a smaller value.

In [14–16], bidirectional switches are added in front of the rectifier in the secondary side as shown in Fig. 7.4a. An alternative way to disconnect the load from the receiving coil is to add a switch after the rectifier [17, 18] as shown in Fig. 7.4b. A diode is also needed to block the reverse current from the output capacitor. The structure is virtually a boost converter without the inductor. The function of the active switch is the same as that of the switches added in front of the rectifier. There are also a few studies on using active rectifiers to realize MEO [19–24]. Figure 7.4c shows the typical circuit topology of such a system. The active rectifier could be constructed by replacing either two off-diagonals or all of the diodes of the full-bridge diode rectifier with active switches. Take the topology shown in Fig. 7.4c as an example. When the lower switches are turned on at the same time, the secondary winding current will circulate in the resonator and will not supply the load.

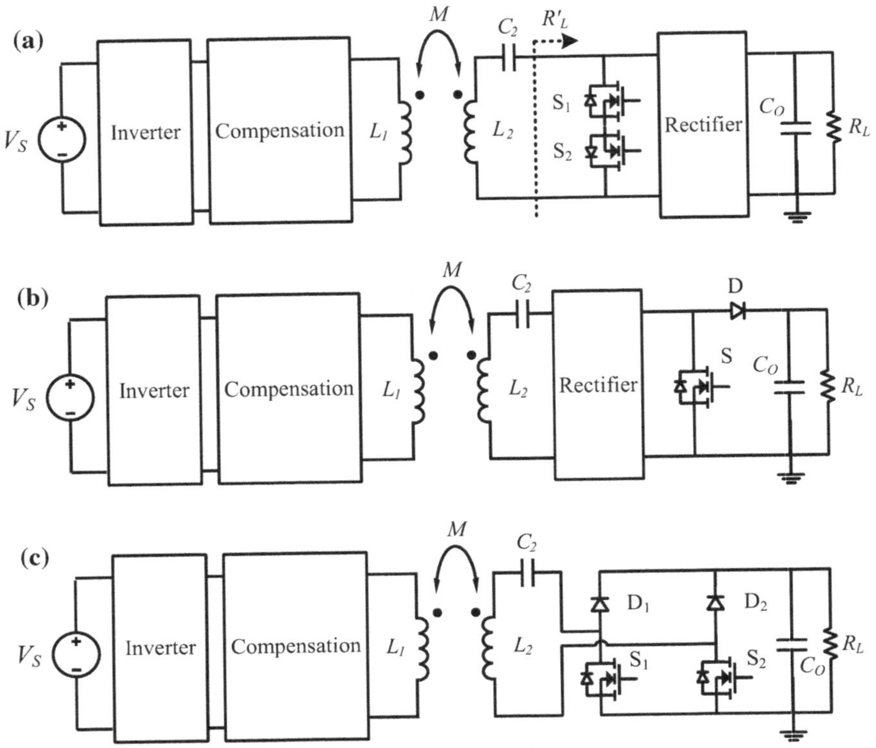


**Fig. 7.4** Three typical boost-type converters for MEO. **a** Switching before rectification, **b** switching after rectification, and **c** switching using an active rectifier

Theoretically, all of the boost-type converters shown in Fig. 7.4 can realize any of the following modulation schemes.

(a) **On–Off Modulation (OFM)**

The switches are controlled to disconnect the load from the receiving coil with a frequency lower than the operating frequency [14, 15]. As a result, the input voltage of the secondary rectifier will appear with the pattern shown in Fig. 7.5a. If a full-bridge rectifier is used, the equivalent load resistance seen by the resonator after adding the active switches is

$$R_L' = (1 - D)^2 \frac{8}{\pi^2} R_L, \tag{7.6}$$

where $D$ is the duty cycle of the active switches. Note that the rectifier converts $R_L$ to $\frac{8}{\pi^2} R_L$. Therefore the active switches bring a factor of $(1 - D)^2$ which is the same as that of a boost converter. However, the derivation of (7.6) is based on the assumption that the amplitude of the current in the secondary winding is unchanged in an on–off

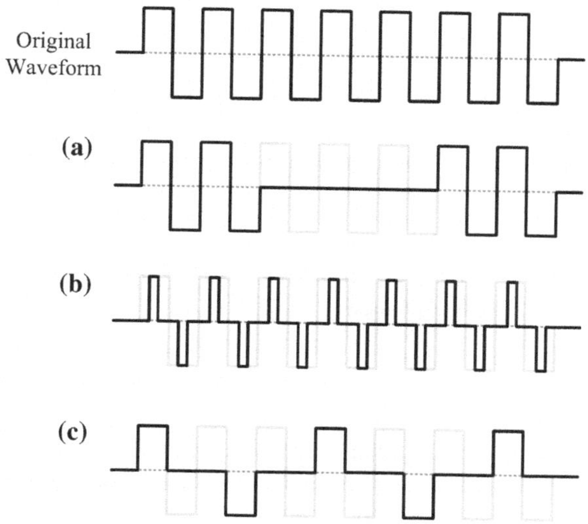


**Fig. 7.5** Input voltages of the secondary rectifier with the typical modulation schemes: **a** on–off modulation; **b** pulse-width modulation; **c** pulse-density modulation

period. The validity of this assumption depends on the frequency of the modulation signal. If the modulation frequency is much lower than the operating frequency (i.e., the switching frequency of the inverter), the amplitude of the secondary winding current cannot be considered constant.

(b) **Pulse-Width Modulation (PWM)**

In [16] and [19–22], the pulse width of the rectifier input voltage is adjusted by controlling the active switches. The pulse width of the rectifier input voltage can be shrunk as shown in Fig. 7.5b, which can be realized by altering the phase-shift between the switching signals of two rectifier legs. The RMS value of the fundamental component of the rectifier input voltage is derived in [19]. Thus, the equivalent load resistance can be derived as

$$R'_L = \frac{8}{\pi^2} R_L \sin\left[(1 - D)\frac{\pi}{2}\right], \tag{7.7}$$

where $D$ is the duty cycle of the active switches. Note that when the active switches are activated, the load is isolated. Therefore, $1 - D$ is corresponding to the pulse width of the rectifier input voltage.

(c) **Pulse-Density Modulation (PDM)**

In [23] and [24], pulse-density modulation methods are adopted for MEO. As shown in Fig. 7.5c, some of the pulses of the rectifier input voltage will be removed according to the ratio between the actual load resistance and the optimal value. It should be noted that the waveform is not necessarily periodic as shown in Fig. 7.5c [24]. It is reported in [24] that the equivalent load resistance after using PDM is

$$R_L' = \frac{8}{\pi^2} R_L (1 - D)^2. \tag{7.8}$$

For consistency, $D$ in (7.8) is also defined as the duty cycle of the active switches and it is corresponding to the time when the load is isolated. Therefore, $1 - D$ is the pulse density of the rectifier input voltage.

### 7.3.3 Using Transmitter-Side On–Off Keying (OOK) Modulation

Reference [25] proposed a transmitter-side OOK modulation which can perform a boost mechanism without using any additional switches or converters. The switching signals of the inverter are modulated with a lower frequency signal and the system operates in on–off mode. The system schematic and the input voltage waveform of the primary resonator (represented with $v_{\text{in}}$) are shown in Fig. 7.6. The equivalent resistance seen into the secondary rectifier is given by

$$R_E = (1 - D)\frac{8}{\pi^2} R_L; 0 < D < 1 \tag{7.9}$$

where $D$ is defined as the ratio between the off time of the inverter over the whole on–off period which is analog to the duty cycle of a boost converter (when the switch is on, the load is not supplied by the source). It should be noted that the factor here is $1 - D$ because the currents are zero during off stage. This is different from boost-type converters using secondary on–off modulation which induces a factor of $(1 - D)^2$. However, transmitter-side OOK modulation is also able to convert the load resistance to a smaller value.

More details of the transmitter-side OOK modulation are included in Chap. 9.

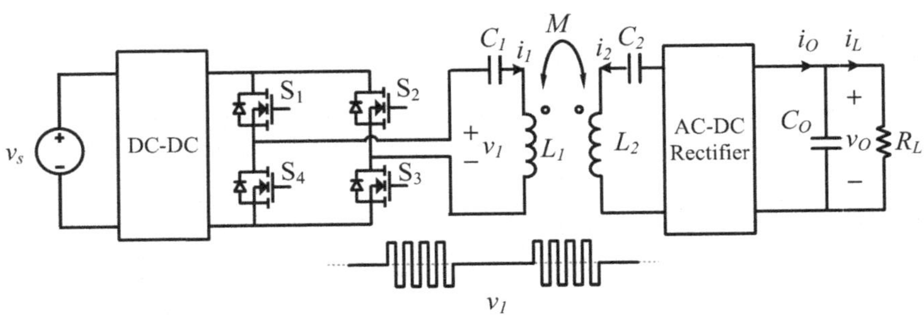


**Fig. 7.6** A WPT system using primary OOK modulation to realize MEO [25]

### 7.3.4 Using Reconfigurable Impedance Transformation Circuits

#### (a) Reconfigurable LC Impedance Matching Networks

LC networks are widely used in RF field for impedance matching. Figure 7.8 shows two typical configurations of impedance matching networks using lump elements (i.e., inductors and capacitors).

Assuming that $R_L$ and $R_{L\_OPT}$ are known, the required $X_1$ in parallel–series connection can be worked out as

$$X_1 = \pm R_L \sqrt{\frac{R_{L\_OPT}}{R_L - R_{L\_OPT}}}. \tag{7.10}$$

From (7.10), we can see that only when $R_L$ is larger than $R_{L\_OPT}$, there is a solution for $X_1$. Therefore, parallel–series connection is not capable of converting a load resistance to a larger value. For series–parallel connection, $R_{L\_OPT}$ is given by

$$R_{L\_OPT} = R_L \frac{X_1^2}{R_L^2 + (X_1 + X_2)^2}. \tag{7.11}$$

By appropriately choosing the values for $X_1$ and $X_2$, $R_{L\_OPT}$ could be either higher or lower than $R_L$.

Using passive impedance matching networks can only transform a load resistance to another fixed value. Therefore it cannot transform a range of load resistance to one specific optimal value. To solve this problem, reconfigurable LC networks have been proposed. In [26], a reconfigurable LC network is used to maintain the resonance of a WPT system against the resonant frequency deviations due to coupling variations. High-frequency relays are used to select the required inductance and capacitance. Although it is not designed to convert the load resistance, the same idea could be applied to realize MEO. In [27], an adaptive optimal load circuit is proposed based on the parallel–series connection shown in Fig. 7.7a. The values of $X_1$ and $X_2$ can be adjusted using switches to connect or disconnect capacitors to the main circuit. As shown in Fig. 7.8, a more flexible design is proposed in [28] using an adjustable

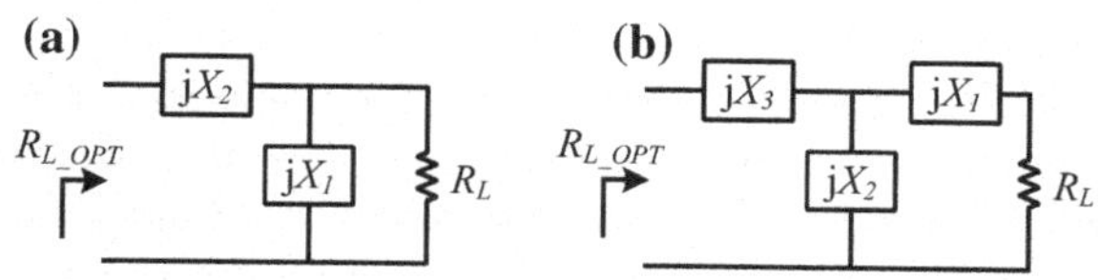


**Fig. 7.7** Two configurations of impedance matching networks. **a** Parallel–series connection: the load is firstly paralleled with reactance $X_1$ then connected in series with $X_2$; **b** series–parallel connection: the load is firstly connected in series with $X_1$ and then paralleled with $X_2$ ($X_3$, if necessary, is used to compensate the reactance of the whole circuit)

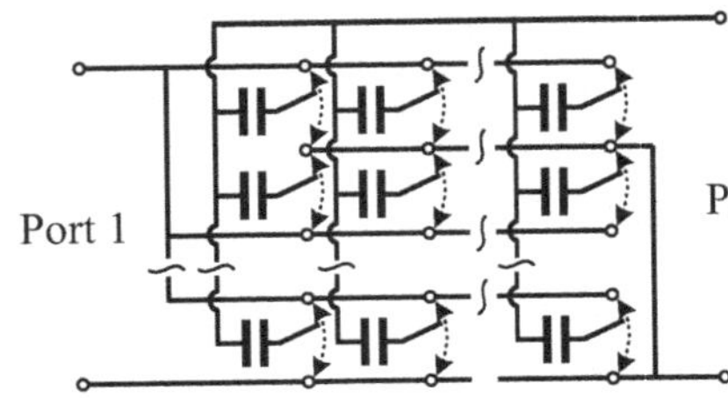


**Fig. 7.8** Capacitor matrix for adaptive impedance matching or load resistance transformation [28]

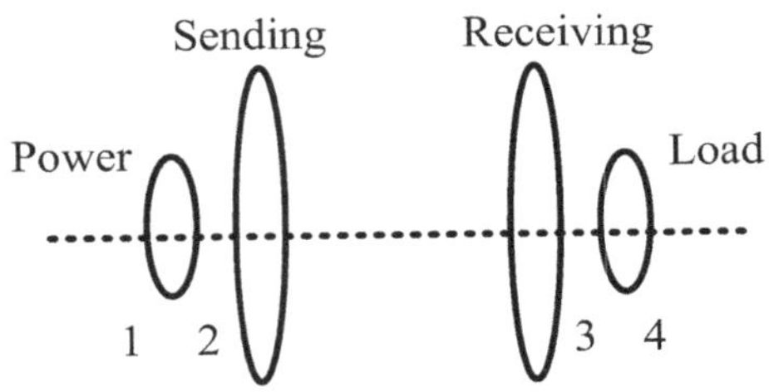


**Fig. 7.9** The four-coil structure for WPT proposed in [30]

capacitor matrix to include more connection possibilities and to reduce the number of switches.

(b) **Adjustable Magnetically Coupled Windings**

The use of magnetically coupled windings and coil-resonator is not new and was reported as early as 1937 in [29], in which a driving coil is magnetically coupled to an antenna (a transmitting coil-resonator) for wirelessly powering a light bulb. This concept was used with coupled windings in both transmitter and receiver sides by MIT's team [30], as shown in Fig. 7.9, the coupling between the **Power** and **Sending** coils and that between the **Receiving** and **Load** coils perform as impedance transformers. Therefore, the **Power** and **Load** coils could be replaced with two lump transformers as reported in [31]. This transformer mechanism was also adopted in the three-resonator structure reported in [32].

A different three-resonator structure with two of the coils in the same plane is studied in [33]. Unlike the analysis of the four-coil structure in which the coupling between the nonadjacent coils is usually neglected, the analysis in [33] includes all the couplings. As a result, the optimal resonant frequencies of the resonators for achieving maximum efficiency are not necessarily equal to the operating frequency of the system.

However, similar to passive impedance matching networks, a pair of coupled windings or a transformer can only convert a single load resistance to the optimal. Hence, reconfigurable winding structures for **Power** and **Load** coils are proposed in [34] and [35] to enlarge the range of load resistance for MEO. In [34], different-diameter loops can be selected for **Power** and **Load** coils as shown in Fig. 7.10a. In [35], the number of turns of the **Power** coil can be adaptively changed as shown in Fig. 7.10b. Both methods can alter the coupling between two coils in the transmitter or receiver side and realize impedance matching for a large range of load resistance.

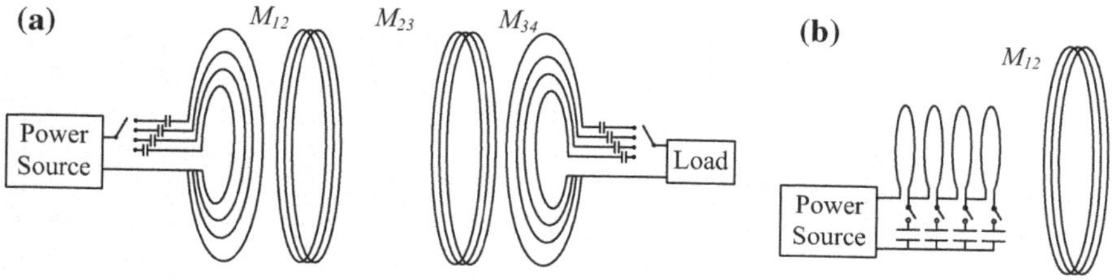


**Fig. 7.10** The reconfigurable winding structures for power and load coils proposed in [34] (**a**) and [36] (**b**)

While in [36], the distance between the **Power** and **Sending** coils and that between the **Receiving** and **Load** coils is adjusted manually for impedance matching for a broader range of power transmission distance.

### 7.3.5 *Using Reconfigurable Coil-Resonant Circuits*

All of the techniques introduced above attempt to convert the load resistance to the optimal value of a given system. Alternatively, the optimal load resistance of a system could also be changed to match the given load resistance. In [37], the receiving coil is multi-tapped as shown in Fig. 7.11. The number of turns of the receiving coil and hence the optimal load resistance of the system can be changed according to the load resistance and coupling coefficient.

Reference [38] combines an impedance matching network and a reconfigurable coil-resonant circuit. The system as shown in Fig. 7.12a is constructed based on a dual-tapped secondary coil. Three switches are used for mode switching. When all the switches are off, it becomes the original two-coil system as shown in Fig. 7.12b. When S1 is off and S2, S3 are on, the load is paralleled with $C_{LP}$ as shown in Fig. 7.12c which can convert the load resistance to a lower value. When S1 is on, and the other two switches are off, the equivalent circuit is shown in Fig. 7.12d. $L_{S1}$ resonates with $C_{S1}$ which forms a repeater. Therefore it changes from a two-coil system to a three-coil system. According to the analysis in [38], the optimal load resistance of the three-coil system could be much lower than that of the original

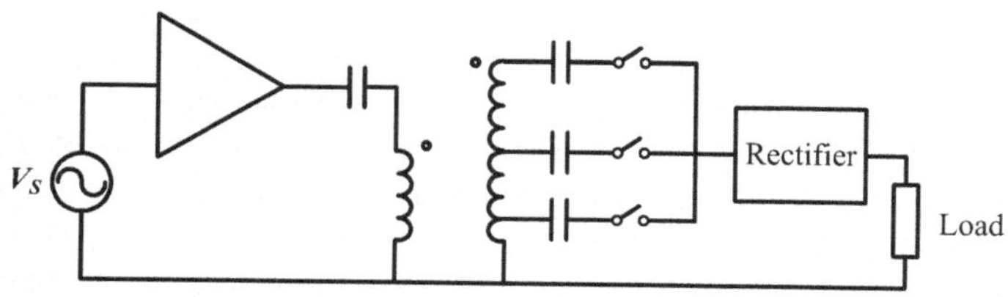


**Fig. 7.11** The multi-tapped receiving coil proposed in [37]

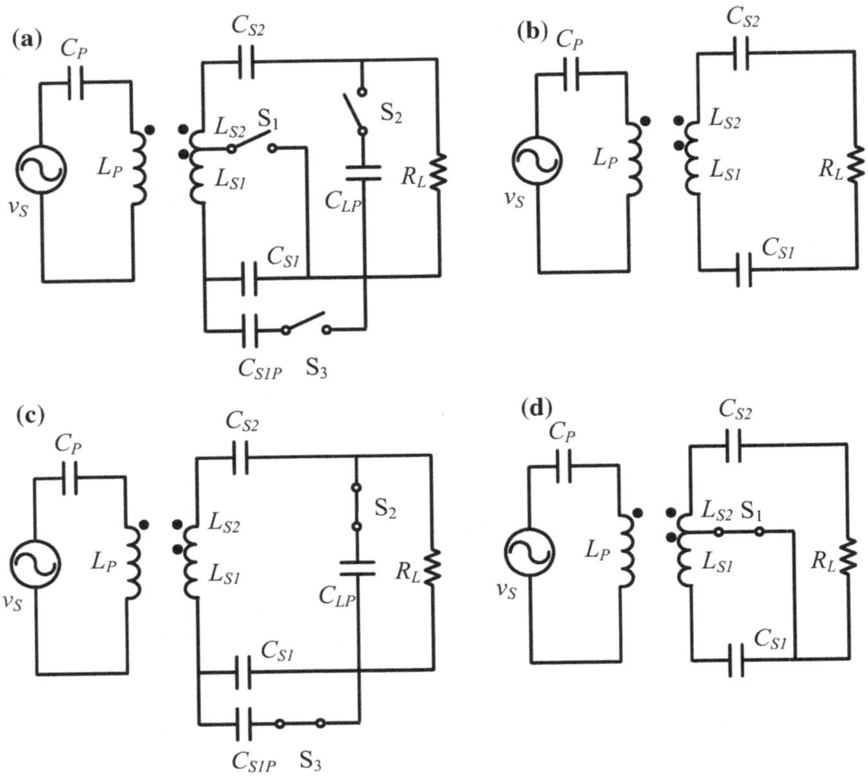


**Fig. 7.12** The proposed WPT system with three reconfigurable topologies in [38] and the equivalent topologies under the three operating modes: **b** two-coil mode, **c** two-coil with parallel capacitor mode, and **d** three-coil mode

two-coil system. By properly choosing the operation mode, system efficiency can be maintained high over a much larger load resistance range compared with that of the original two-resonator system. Figure 7.13 shows the efficiency performance.

## 7.4 Review of MEO Control Schemes

MEO techniques aim at meeting the condition that *the load resistance equals the optimal load resistance* for a WPT system to achieve the highest possible efficiency. Apart from choosing one of the above methods, it is also essential to locate and determine the optimal value for the control variable (such as the duty cycle in most of the cases) to achieve MEO. Generally, there are four control schemes that can locate the optimal control variable.

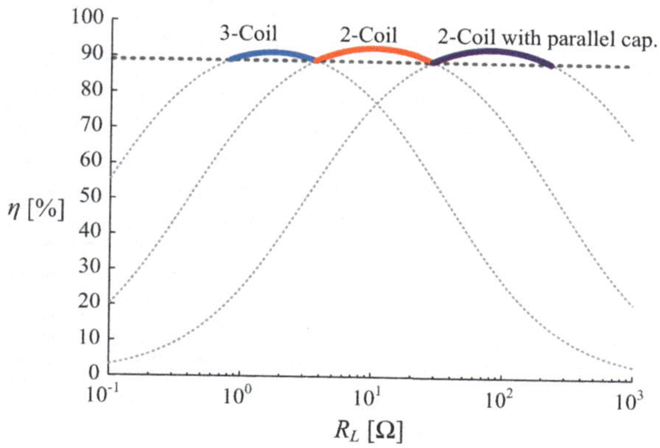


**Fig. 7.13** Efficiency performance of the reconfigurable system proposed in [38]

### *7.4.1 Perturbation and Observation (P&O)*

Ideally, both the actual load resistance and the optimal load resistance of a WPT system should be known to convert one of them to the other. Load resistance can be determined by sensing the load voltage and load current. But the optimal load resistance of a WPT system might change as the mutual coupling changes from time to time or even keeps changing, for example, in a dynamic EV charging application. Obtaining the precise mutual coupling of a WPT system is difficult. P&O is a well-known maximum power point tracking (MPPT) method used in photovoltaic (PV) converters. P&O can locate the maximum-efficiency point without the information of the load and mutual coupling. Instead of directly calculating the optimal value for the control variable, system efficiency is monitored under a series of perturbations, and the maximum-efficiency point can be eventually found. There are two scenarios in the implementation of P&O, i.e., with [5, 6] or without [1, 7] primary–secondary communication.

(a) **With Primary–Secondary Communication**

Figure 7.14a shows the flow chart of the P&O tracking scheme with a primary–secondary communication channel. $D_n$ represents the control variable at the present instant. For example, when using a DC–DC converter, $D_n$ is the duty cycle of the converter. In each P&O step, $D_n$ will be changed a little bit, and then the system efficiency will be measured after the system reaches a steady state. The efficiency will be compared with the last measurement, and it can be determined whether $D_n$ should be further changed in the same direction or changed in the reversed direction. P&O tracking will continue until the maximum-efficiency point is found.

(b) **Without Primary–Secondary Communication**

P&O tracking can also be realized without primary–secondary communication. In [1], phase-shift control is applied to the primary inverter to adjust the input power of the system, while the secondary buck–boost converter is regulating the output

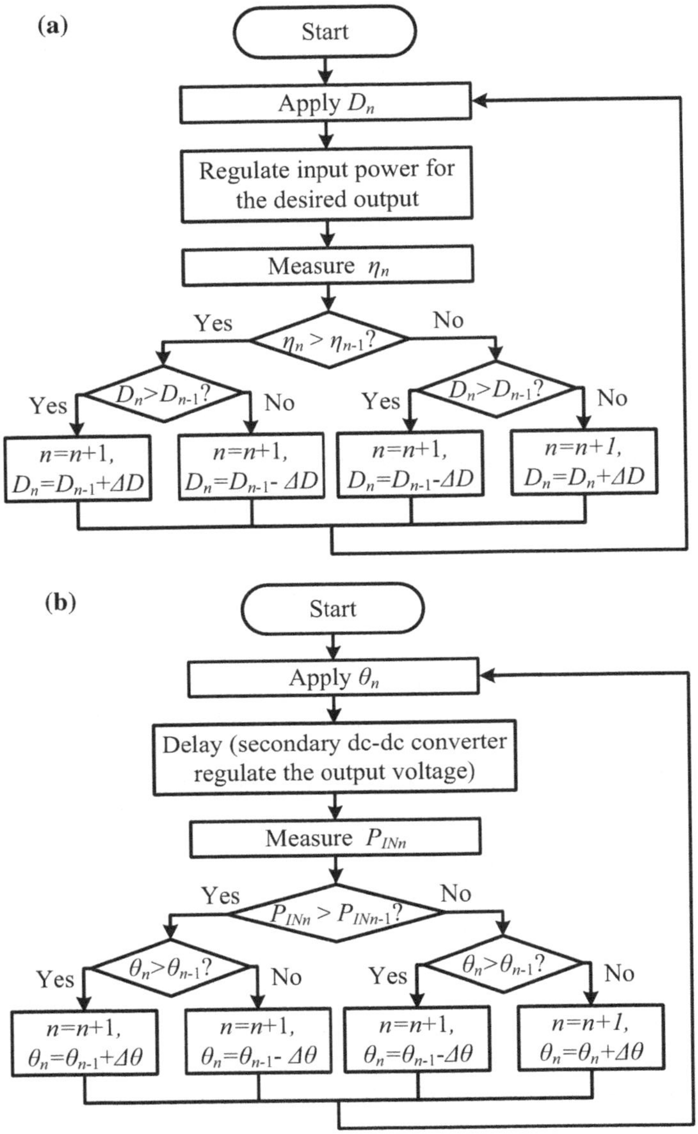


**Fig. 7.14** Flowchart of the P&O tracking scheme: **a** with a communication channel, **b** without a communication channel

voltage to a desired constant value. Thereby, the buck–boost converter could use a simple PI controller. The flow chart is shown in Fig. 7.14b where $\theta$ represents the phase-shift of the primary inverter. After applying a new phase-shift to the inverter, there will be a delay before measuring the input power, because it will take some time for the DC–DC converter to reach a new steady state. If the output power can be considered as constant in a short period which is the case for battery charging, the maximum efficiency of the system can be found by searching for the minimum input power. Typically, the DC voltage fed to the inverter is constant. Thus minimum input current of the inverter will be equivalent to minimum input power. It has been reported in [1] that the relationship between the output voltage of the system and the duty cycle of the buck–boost converter might be contrary to the voltage gain characteristic of a buck–boost converter fed with a voltage source. Adopting the basic idea of P&O tracking without primary–secondary communication, paper [7] reported that both phase-shift and operating frequency could be regulated to push up the system efficiency.

P&O tracking method does not need the information of the coupling. However, the searching time might become an issue when the coupling or the output power is changing rapidly in applications like EV dynamic charging.

### 7.4.2 *Calculating Optimal Control Variable Based on Coupling Estimation*

The optimal load resistance depends on the winding resistances and the mutual inductance (or self-inductance and coupling coefficient) of the coils. In most of the applications, winding resistances and self-inductances can be considered as constants regardless of the variation in the relative positions of the coils. Hence the coils will maintain resonant, and only the coupling coefficient is unknown. If coupling coefficient or mutual inductance can be calculated with some of the voltages and/or currents in the system, the optimal load resistance, as well as the optimal duty cycle for the converters (or switches, or modulation signal) can be worked out accordingly.

It was reported in [8] that mutual inductance can be estimated using information from either the primary side or secondary side. From the primary side, the input impedance seen into the primary resonator can be calculated with the aid of the basic circuit equations of the system. The real part and imaginary part of the input impedance are given by

$$\mathrm{Re}(Z_{\mathrm{in}}) = R_{P1} + \frac{\omega^2 M^2 (Z_L + R_{P2})}{(Z_L + R_{P2})^2 + X_2^2} \tag{7.12}$$

$$\mathrm{Im}(Z_{\mathrm{in}}) = X_1 - \frac{\omega^2 M^2 X_2}{(Z_L + R_{P2})^2 + X_2^2}. \tag{7.13}$$

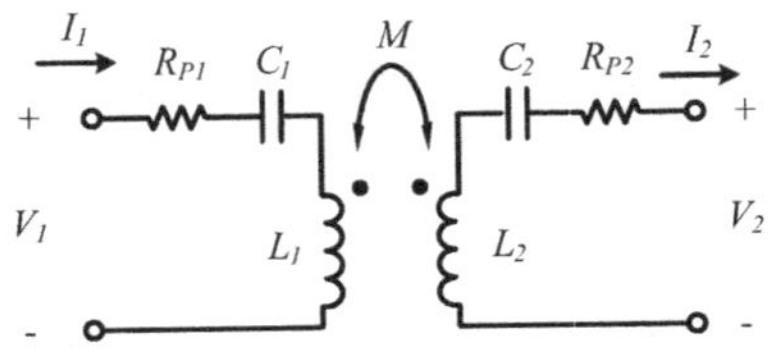


**Fig. 7.15** Symbols of a WPT system for coupling calculation

By solving (7.12), $M$ can be obtained as

$$M = \frac{1}{\omega}\sqrt{\frac{[\mathrm{Re}(Z_{\mathrm{in}}) - R_{P1}][(Z_L + R_{P2})^2 + X_2^2]}{Z_L + R_{P2}}}. \tag{7.14}$$

If $X_2 \neq 0$ at the frequency used for coupling estimation, $M$ could also be calculated with (7.13).

In (7.14), $R_{P1}$, $R_{P2}$, $Z_L$, and $X_2$ are known parameters. Re($Z_{\mathrm{in}}$) can be calculated with the sensed input voltage and input current of the primary resonator. However, in real situations, $Z_L$ might vary with time. Therefore, $Z_L$ must be calculated using the sensed load voltage and current, and then it will be transmitted to the primary-side controller.

By using the secondary-side information, the mutual inductance can also be worked out [9] as

$$M = \frac{V_1 \pm \sqrt{V_1^2 - 4R_{P1}I_2(V_2 + R_{P2}I_2)}}{2I_2\omega}, \tag{7.15}$$

where, as shown in Fig. 7.15, $V_1$ is the RMS value of the fundamental component of the input voltage applied to the primary resonator, $V_2$ and $I_2$ are the RMS values of the fundamental components of the voltage and current output by the secondary resonator, respectively. When using a full-bridge inverter to supply the primary resonator, $V_1$ can be calculated with the DC source voltage $V_S$, as

$$V_1 = \frac{2\sqrt{2}}{\pi}V_S. \tag{7.16}$$

If a full-bridge rectifier is used to supply the load, $I_2$ can be calculated with the load current $I_L$, as

$$I_2 = \frac{\sqrt{2}\pi}{4}I_L. \tag{7.17}$$

Equation (7.15) shows that there are two solutions for mutual inductance when calculating it without primary-side information. Therefore, we need two different operation points (i.e., two sets of parameters) to determine the correct mutual inductance. In [8] and [10], two different duty cycles of the secondary converter were used to obtain two different operation points.

Based on the same equation of (7.15), a dynamic coupling estimation method is proposed in [10]. In a short period, the coupling coefficient is considered to change linearly with respect to the duty cycle of the primary DC–DC converter. Therefore, the latest coupling coefficient can be calculated with the estimated coupling coefficient in the last estimation period and the duty cycle change of the primary DC–DC converter. It is believed that this method can be applied to estimate the coupling of a dynamic charging system. For stationary charging, it is not necessary to keep tracking the coupling since it will not change in the charging process. One of the issues of directly calculating the control variable for MEO is that (7.15) is derived based on fundamental components of the waveforms. Higher order harmonics will affect the accuracy of coupling estimation.

In [39], a method is proposed to estimate any parameter including mutual inductance of a WPT system by using only the measured input voltage and current in the primary side. The input impedance of the system at multiple frequencies is calculated according to the measured input voltage and current. Then the Genetic Algorithm (GA) is used to locate the exact values for the unknown parameters.

### *7.4.3 Voltage Ratio Control*

In [19], the maximum-efficiency point is achieved by controlling the ratio between $V_1$ and $V_2$ to a constant at which MEO can be realized. $V_1$ and $V_2$ are the voltages as shown in Fig. 7.15. A similar strategy is adopted by [13] in which the reason of using constant voltage ratio is discussed. When $\frac{\omega^2 M^2}{R_{P1} R_{P2}} \gg 1$ and $\frac{\omega^2 M^2}{R_{P1} R_L} \gg 1$, the voltage ratio is given by

$$\frac{V_2}{V_1} = \frac{\omega M}{\omega^2 M^2 + R_{P1}(R_{P2} + R_L)} R_L \approx \frac{1}{\omega M} R_L. \tag{7.18}$$

Therefore, the optimal voltage ratio can be worked out with the optimal load resistance given by (2.24).

$$\left(\frac{V_2}{V_1}\right)_{\text{OPT}} \approx \frac{1}{\omega M} R_{L\_\text{OPT}} = \frac{1}{\omega M} R_{P2} \sqrt{1 + \frac{\omega^2 M^2}{R_{P1} R_{P2}}} \approx \sqrt{\frac{R_{P2}}{R_{P1}}}. \tag{7.19}$$

Equation (7.19) shows that the optimal voltage ratio is determined by the ratio of the resistance of the coils and independent of the coupling and the load resistance [13].

## 7.5 Comparison and Discussion

Table 7.1 summarizes and compares the basic aspects of the existing MEO techniques. From cost and complexity point of view, using standard DC–DC converters is likely the worst solution, and transmitter-side OOK modulation is the best since it does not need any extra components.

### 7.5.1 Light-Load Conditions ($R_L > R_{L\_OPT}$)

In many studies, one assumption is that a WPT system has been properly designed so that the inherent optimal load resistance is equal to the actual load resistance at the rated power. For a system with a constant output voltage, the load resistance at the rated power is also the minimum load resistance. Therefore, the system operates with the highest possible efficiency at the rated power. The only concern is the efficiency under light-load conditions (normally refers to $R_L > R_{L\_OPT}$). In this case, using boost-type converters or transmitter-side OOK modulation is simpler and more efficient compared with using standard DC–DC converters or using reconfigurable impedance transformation circuits. Compared with boost-type converters, transmitter-side OOK modulation not only uses no additional active switches or components but also minimizes the rectification loss [25]. But it generates larger output voltage ripples due to the lower frequency modulation signal. Among the three boost-type converters, an active rectifier may be the best choice since it has the lowest component count.

**Table 7.1** Comparison of MEO techniques

| | Increased cost | Increased complexity | Soft switching | $R_L < R_{L\_OPT}$ | $R_L > R_{L\_OPT}$ | Smooth conversion | Output voltage ripple |
|---|---|---|---|---|---|---|---|
| Standard DC–DC converters | High | High | Yes | Yes | Yes | Yes | Small |
| Boost-type converters | Medium | Medium | Depends | Yes | No | Yes | Depends |
| Transmitter-Side OOK modulation | Low | Low | In part | Yes | No | Yes | Large |
| Reconfigurable impedance transformation circuits | Medium | High | Yes | Yes | Yes | No | Small |
| Reconfigurable coil-resonant circuits | Low | Medium | Yes | Yes | Yes | No | Small |

Then among the modulation methods of boost-type converters, OFM has the largest output voltage ripple, and PDM is medium, i.e., OFM > PDM > PWM with respect to output voltage ripple. On the other hand, PWM is not able to realize soft switching unless introducing reactive power in the secondary side [40, 41]. PDM may have larger harmonics when a large number of pulses have been removed under light-load conditions.

### 7.5.2 For Arbitrary Load Resistance

It has been pointed out that maximizing efficiency is equivalent to maximizing the coupling coefficient and the quality factors. Therefore, in a real application, the optimal load resistance is determined according to (2.24) after efficiency optimization. This optimal load resistance might be larger than the actual equivalent load resistance at rated power, and thus it is necessary to convert the load resistance to either a larger or a smaller value, depending on the load power of the system. However, most of the existing techniques which can perform arbitrary resistance transformation have apparent disadvantages. (i) Using DC–DC converters will increase the complexity and cost of the system, and will also cause extra losses. (ii) Reconfigurable impedance transformation circuits use a lot of switches and passive components which increase the complexity and size of the system. Reconfigurable coil-resonant circuits demonstrate a new perspective by changing the inherent optimal load resistance of a WPT system, and it may provide a promising solution when it is combined with other techniques.

### 7.5.3 Control Schemes Comparison

P&O tracking can locate the maximum-efficiency point regardless of any parameter variations including coupling, load resistance, inductance, capacitance, etc. However, P&O suffers from long response time, and thus it is not likely to be applied to dynamic charging applications with a fast-changing coupling. But P&O is entirely competent in stationary wireless charging applications where the coupling will not change in the charging process and the load resistance of the battery will vary slowly. For dynamic charging, voltage ratio control method is better than directly calculating the control variable based on coupling estimation, because it does not need to sense the currents and has higher accuracy. The constraint of voltage ratio control is that the tolerance of the inductance and capacitance should be carefully evaluated.

## 7.6 Conclusion

This chapter provides a review on the MEO techniques. For improving the efficiency of light-load operation, transmitter-side OOK modulation offers the best efficiency performance but will have larger output voltage ripples; secondary active rectifier with either PWM or PDM is also cost-effective with two active switches and the corresponding control circuits. For arbitrary load resistance transformation, a reconfigurable coil-resonant circuit combined with impedance transformation circuit provides a promising solution expect that continuous conversion is not realized.

## References

1. Zhong WX, Hui SYR (2015) Maximum energy efficiency tracking for wireless power transfer systems. IEEE Trans Power Electron 30(7):4025–4034
2. Hui SYR, Zhong W, Lee CK (2014) A critical review of recent progress in mid-range wireless power transfer. IEEE Trans Power Electron 29(9):4500–4511
3. Bosshard R, Kolar JW, Mühlethaler J, Stevanović I, Wunsch B, Canales F (2015) Modeling and $\eta$-$\alpha$-Pareto optimization of inductive power transfer coils for electric vehicles. IEEE J Emerg Sel Top Power Electron 3(1):50–64
4. Moriwaki Y, Imura T, Hori Y (2011) Basic study on reduction of reflected power using DC/DC converters in wireless power transfer system via magnetic resonant coupling. In: 2011 IEEE 33rd international telecommunications energy conference (INTELEC), Amsterdam, pp 1–5
5. Fu M, Yin H, Zhu X, Ma C (2015) Analysis and tracking of optimal load in wireless power transfer systems. IEEE Trans Power Electron 30(7):3952–3963
6. Li H, Li J, Wang K, Chen W, Yang X (2015) A maximum efficiency point tracking control scheme for wireless power transfer systems using magnetic resonant coupling. IEEE Trans Power Electron 30(7):3998–4008
7. Patil D, Sirico M, Gu L, Fahimi B (2016) Maximum efficiency tracking in wireless power transfer for battery charger: Phase shift and frequency control. In: 2016 IEEE energy conversion congress and exposition (ECCE), Milwaukee, WI, pp 1–8
8. Jiwariyavej V, Imura T, Hori Y (2015) Coupling coefficients estimation of wireless power transfer system via magnetic resonance coupling using information from either side of the system. IEEE J Emerg Sel Top Power Electron 3(1):191–200
9. Kobayashi D, Imura T, Hori Y (2015) Real-time coupling coefficient estimation and maximum efficiency control on dynamic wireless power transfer using secondary DC-DC converter. In: IECON 2015—41st annual conference of the IEEE Industrial Electronics Society, Yokohama, 2015, pp 4650–4655
10. Dai X, Li X, Li Y, Hu AP (2018) Maximum efficiency tracking for wireless power transfer systems with dynamic coupling coefficient estimation. IEEE Trans Power Electron 33(6):5005–5015
11. Fu M, Yin H, Liu M, Ma C (2016) Loading and power control for a high-efficiency Class E PA-driven megahertz WPT system. IEEE Trans Ind Electron 63(11):6867–6876
12. Yeo TD, Kwon D, Khang ST, Yu JW (2017) Design of maximum efficiency tracking control scheme for closed-loop wireless power charging system employing series resonant tank. IEEE Trans Power Electron 32(1):471–478
13. Huang Z, Wong SC, Tse CK (2018) Control design for optimizing efficiency in inductive power transfer systems. IEEE Trans Power Electron 33(5):4523–4534
14. Ahn D, Kim S, Moon J, Cho IK (2016) Wireless power transfer with automatic feedback control of load resistance transformation. IEEE Trans Power Electron 31(11):7876–7886

15. Tang X, Zeng J, Pun KP, Mai S, Zhang C, Wang Z (2018) Low-cost maximum efficiency tracking method for wireless power transfer systems. IEEE Trans Power Electron 33(6):5317–5329
16. Kiani M, Lee B, Yeon P, Ghovanloo M (2015) A Q-modulation technique for efficient inductive power transmission. IEEE J Solid-State Circuits 50(12):2839–2848
17. Yuan L, Li B, Zhang Y, He F, Chen K, Zhao Z (2015) Maximum efficiency point tracking of the wireless power transfer system for the battery charging in electric vehicles. In: 2015 18th international conference on electrical machines and systems (ICEMS), Pattaya, pp 1101–1107
18. Zhong W, Hui SYR (2017) Charging time control of wireless power transfer systems without using mutual coupling information and wireless communication system. IEEE Trans Ind Electron 64(1):228–235
19. Diekhans T, De Doncker RW (2015) A dual-side controlled inductive power transfer system optimized for large coupling factor variations and partial load. IEEE Trans Power Electron 30(11):6320–6328
20. Berger A, Agostinelli M, Vesti S, Oliver JA, Cobos JA, Huemer M (2015) A wireless charging system applying phase-shift and amplitude control to maximize efficiency and extractable power. IEEE Trans Power Electron 30(11):6338–6348
21. Colak K, Asa E, Bojarski M, Czarkowski D, Onar OC (2015) A novel phase-shift control of semibridgeless active rectifier for wireless power transfer. IEEE Trans Power Electron 30(11):6288–6297
22. Mai R, Liu Y, Li Y, Yue P, Cao G, He Z. An active rectifier based maximum efficiency tracking method using an additional measurement coil for wireless power transfer. IEEE Trans Power Electron (in press)
23. Li H, Fang J, Chen S, Wang K, Tang Y. Pulse density modulation for maximum efficiency point tracking of wireless power transfer systems. IEEE Trans Power Electron
24. Li H, Wang K, Fang J, Tang Y. Pulse density modulated ZVS full-bridge converters for wireless power transfer systems. IEEE Trans Power Electron (in press)
25. Zhong W, Hui SYR (2018) Maximum energy efficiency operation of series-series resonant wireless power transfer systems using on-off keying modulation. IEEE Trans Power Electron 33(4):3595–3603
26. Beh TC, Kato M, Imura T, Oh S, Hori Y (2013) Automated impedance matching system for robust wireless power transfer via magnetic resonance coupling. IEEE Trans Ind Electron 60(9):3689–3698
27. Jung YK, Lee B (2013) Design of adaptive optimal load circuit for maximum wireless power transfer efficiency. In: 2013 Asia-Pacific microwave conference proceedings (APMC), Seoul, pp 1221–1223
28. Lim Y, Tang H, Lim S, Park J (2014) An adaptive impedance-matching network based on a novel capacitor matrix for wireless power transfer. IEEE Trans Power Electron 29(8):4403–4413
29. Gernsback H (ed) Lighting lamp by S-W-radio. Short wave & television, pp 166 and 191
30. Kurs A, Karalis A, Moffatt R, Joannopoulos JD, Fisher P, Soljacic M (2007) Wireless power transfer via strongly coupled magnetic resonances. Sci Express 317:83–86
31. Choi BH, Lee ES, Huh J, Rim CT (2015) Lumped impedance transformers for compact and robust coupled magnetic resonance systems. IEEE Trans Power Electron 30(11):6046–6056
32. Kiani M, Jow U, Ghovanloo M (2011) Design and optimization of a three-resonator inductive link for efficient wireless power transmission. IEEE Trans Biomed Circuits Syst 5:579–591
33. Zhong WX, Zhang C, Liu X, Hui SYR (2015) A methodology for making a three-resonator wireless power transfer system more energy efficient than a two-resonator counterpart for extended transmission distance. IEEE Trans Power Electron 30(2):933–942
34. Kim J, Choi W-S, Jeong J (2013) Loop switching technique for wireless power transfer using magnetic resonance coupling. Progr Electromagn Res 138:197–209
35. Lee G, Waters BH, Shin YG, Smith JR, Park WS (2016) A reconfigurable resonant coil for range adaptation wireless power transfer. IEEE Trans Microw Theory Tech 64(2):624–632
36. Duong TP, Lee JW (2011) Experimental results of high-efficiency resonant coupling wireless power transfer using a variable coupling method. IEEE Microw Wirel Compon Lett 21(8):442–444

37. Mercier PP, Chandrakasan AP (2013) Rapid wireless capacitor charging using a multi-tapped inductively-coupled secondary coil. IEEE Trans Circuits Syst I Regul Pap 60(9):2263–2272
38. Zhong W, Hui SYR. Reconfigurable wireless power transfer systems with high energy efficiency over wide load range. IEEE Trans Power Electron (in press)
39. Lin D, Yin J, Hui SYR (2014) Parameter identification of wireless power transfer systems using input voltage and current. In: 2014 IEEE energy conversion congress and exposition (ECCE), Pittsburgh, PA, pp 832–836
40. Liu F, Lei W, Wang T, Nie C, Wang Y (2017) A phase-shift soft-switching control strategy for dual active wireless power transfer system. In: 2017 IEEE energy conversion congress and exposition (ECCE), Cincinnati, OH, pp 2573–2578
41. Jiang Y, Wang L, Wang Y, Liu J, Li X, Ning G. Analysis, design and implementation of accurate ZVS angle control for EV's battery charging in wireless high power transfer. IEEE Trans Ind Electron (in press)

# Chapter 8
# Using a DC–DC Converter and the P&O Scheme for MEO Without Transmitter and Receiver Communication—A Design Example

## 8.1 Introduction

For solar power applications, maximum power point tracking (MPPT) is a well-known concept. A similar idea could be applied to WPT systems, except that the objective is now to maximize the efficiency for WPT. Since a review on MEO techniques and the related references has been given in Chap. 7, the coming chapters will not include any more literature review or references.

In this chapter, the method of using a DC–DC converter and the P&O control scheme for MEO without transmitter and receiver communication is explored and evaluated. Using the DC–DC converter in the receiver module to emulate the optimal equivalent load condition dynamically, the method ensures automatic MEO by searching for the minimum input power operation point for a given output power. The operating principle is demonstrated in a two-resonator WPT system.

## 8.2 Searching for the Optimal Duty Cycle

In this approach, a buck–boost converter is used to regulate the output and maintain a constant output voltage, as shown in Fig. 8.1. The transmitter power circuit initially provides an input voltage of sufficiently high value to produce the desired output voltage. Such input voltage $v_{\text{in}}$ of the primary resonator is then decreased iteratively by adopting either a DC–DC converter in front of the inverter or a phase-shift control for the full-bridge inverter, until the input power of the transmitter circuit reaches a minimum point. For a given output power, reaching a minimum input power operation point is equivalent to reaching the maximum efficiency operation point. This approach is based on the assumption that the load power changes much slower than this iterative searching process. This assumption is valid for battery charging applications.



W. Zhong et al., *Wireless Power Transfer*, CPSS Power Electronics Series,
https://doi.org/10.1007/978-981-15-2441-7_8

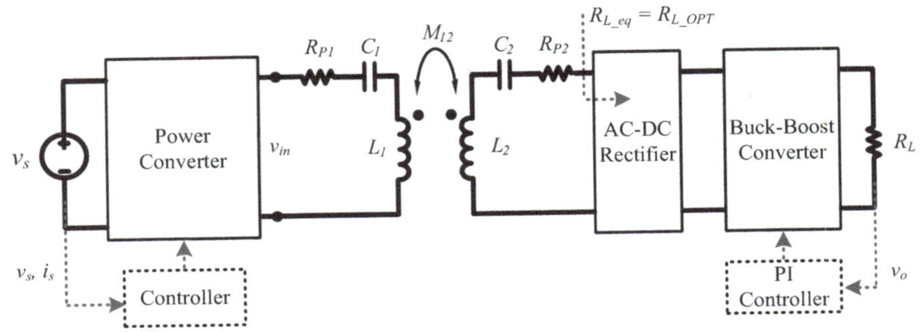


**Fig. 8.1** Schematic of a two-resonator WPT system with a buck–boost converter in the secondary side and without transmitter and receiver communication

The flow chart of this approach is shown in Fig. 7.14b and explained in the last chapter.

The design example is based on a WPT system with two identical coils as shown in Figs. 8.2 and 8.3. The parameters of the system are specified in Table 8.1.

**Fig. 8.2** The coil of a practical WPT system

**Fig. 8.3** A WPT system with a coupling coefficient of about 0.1

**Table 8.1** Parameters of the practical WPT system

| Parameters | Symbol | Practical value |
|---|---|---|
| Inner diameter | $d_i$ | 21.7 mm |
| Coil width | $W_r$ | 5.32 mm |
| Outer diameter | $d_o$ | 27.02 mm |
| Number of turns per layer | – | 9 |
| Number of layers | – | 2 |
| Self-inductance of the transmitter | $L_1$ | 23.2 μH |
| Self-inductance of the receiver | $L_2$ | 24.0 μH |
| Mutual inductance | $M_{12}$ | 2.56 μH |
| Compensating capacitance of the transmitter | $C_1$ | 106.2 nF (theoretical: 109.2 nF) |
| Compensating capacitance of the receiver | $C_2$ | 105.3 nF (theoretical 105.5 nF) |
| Resistance of each coil | $R_{P1}$ and $R_{P2}$ | 0.3 Ω |

The operating frequency of the power inverter in the transmitter circuit is set at 100 kHz which is approximately equal to the resonance frequencies of the two resonators. The litz wire has 24 strands of 0.08 mm strand diameter and a ferrite plate with a thickness of 1 mm is used for enhancing the mutual coupling.

In the case study, the desired output voltage of the system is 2.5 V and the rated output power is 2.5 W. Therefore, the load resistance can vary in the range from 2.5 Ω to $+\infty$. Therefore, the relationship between the input resistance $R_{L\_eq}$ of the buck–boost converter and the load resistance $R_L$ is given by (7.5).

For the rated load resistance $R_L = 2.5\ \Omega$, the relationship between the duty cycle $D$ of the buck–boost converter and the efficiency of the system (assuming all the converters are lossless) and the input voltage of the primary resonator (i.e., the output voltage of the power inverter) is shown in Fig. 8.4. The searching process of the maximum energy efficiency operation point is now illustrated with the aid of Fig. 8.4. The purple solid line in Fig. 8.4 represents the required input voltage of the power inverter on the transmitter side in order to generate an output voltage of 2.5 V on the receiver side for a load resistance of 2.5 Ω. At $D = 0.38$, where is indicated with a vertical dotted line, the input voltage required for an output power of 2.5 W in this example is at its minimum value. The dotted line separates the system operating range into two regions. The relationships between the duty cycle and the output voltage (or output power) are opposite in these two regions. On the left side ($D < 0.38$), the output power increases as $D$ increases, thereby a lower input voltage is required when $D$ becomes larger. On the right side ($D > 0.38$), the output power decreases as $D$ increases, thereby a higher input voltage is required when $D$ becomes larger. It is important to note that, only the region where the maximum efficiency point exists should be used for MEO. In this example, the region on the right side of the dotted line should be chosen as the operating region. If 10 V is used as the initial input voltage, the searching process will allow the input voltage to change gradually

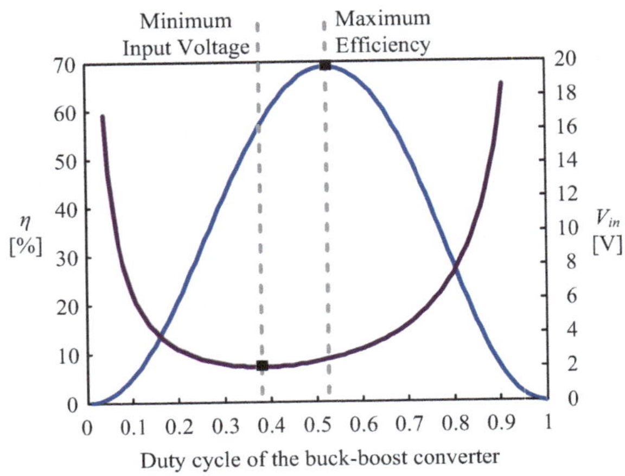


**Fig. 8.4** Efficiency of the system and the required input voltage as $D$ varies

to about 2.5 V at which the efficiency is at its maximum of 69%. In Fig. 8.5, the efficiency is plotted against the input voltage of the primary resonator. One method to restrict the operation of the system within the right region is to restrict the duty cycle to be larger than 0.38.

However, on the right region of the dotted line in Fig. 8.4, the output DC voltage of the buck–boost converter increases as $D$ decreases. This output voltage and duty cycle relationship is opposite to the voltage gain characteristic of a standard buck–boost converter, i.e., the output voltage increases with increasing duty cycle. One method to implement the required control characteristics is to use the complement of the duty cycle, i.e., $D' = 1 - D$ as shown in Fig. 8.6. This can be practically achieved by using an inverter gate to invert the gate signal for the buck–boost converter.

Alternatively, a standard buck–boost converter could be used without using the complement of the duty cycle if one can adjust the minimum input voltage point to

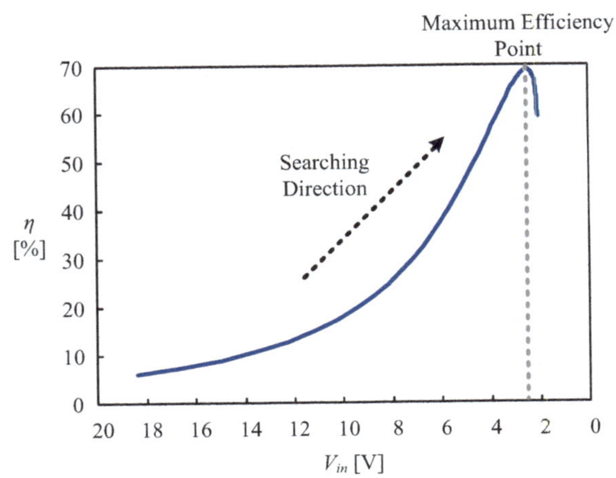


**Fig. 8.5** Efficiency variation as the input voltage decreases and the duty cycle of the buck–boost converter is forced to search for a value to output the designated voltage

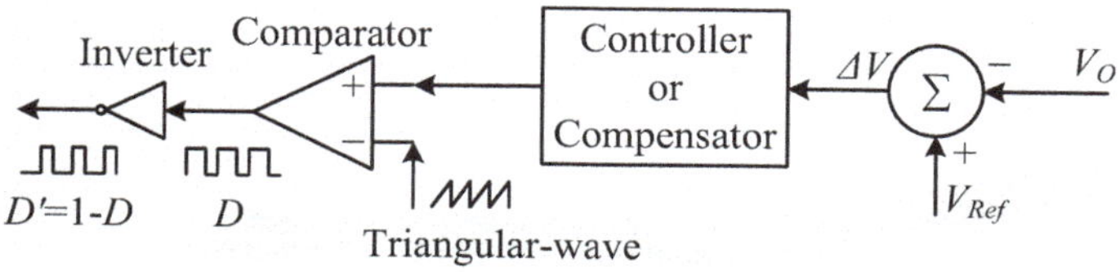


**Fig. 8.6** Control loop for a buck–boost converter operating on the right region

the left of the maximum efficiency point. This can be done by (i) tuning the operating frequency or (ii) tuning the natural resonant frequency of the transmitter by adjusting the value of the compensation capacitor ($C_1$). However, for the first method (i), the maximum efficiency of the system will be degraded due to nonzero impedance $X_2$. For the second method (ii), the VA requirement of the power source increases due to nonzero $X_1$.

## 8.3 Experimental Verifications

For the case study based on parameters in Table 8.1, experiments have been carried out to evaluate the performance of the two-resonator WPT system analyzed above. A full-bridge power inverter is used in the transmitter circuit to generate a 100 kHz AC supply for the transmitting coil. The DC input of the inverter is fixed at 5 V. The output voltage of the inverter is adjusted by applying phase-shift control. For a load resistance of 2.5 Ω and an output voltage of 2.5 V, Fig. 8.7a and b shows the output voltages of the inverter at the initial point and the optimal point, respectively. The pulse width decreases from the original 5–2.26 μs when the optimal point is located. In Fig. 8.8, the DC source current of the system is shown. Since the DC source voltage is constant, the source current is proportional to the input power of the system. As indicated in Fig. 8.8, the source current reaches the minimum point at

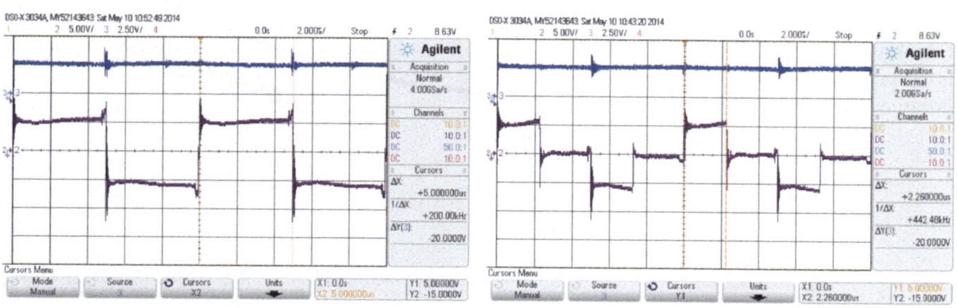


(a) at the initial point (b) at the optimal point

**Fig. 8.7** The input voltage (bottom) of the primary coil and the output voltage (top) of the system when $R_L = 2.5\ \Omega$

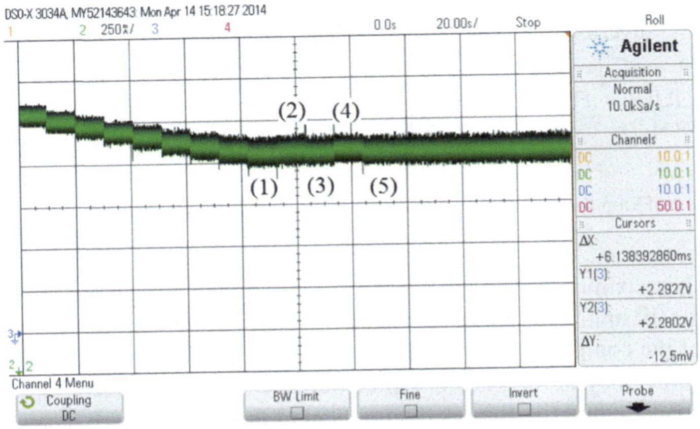


**Fig. 8.8** The source current of the system

(1) when the pulse width equals to 2.26 μs; and the searching process continues and the pulse width is decreased to 2.135 μs; then the source current becomes larger at (2); so the duty cycle will be changed back to 2.26 μs at (3). Normally, the searching process will stop at (3) because the optimal point has been found. However, in order to eliminate the errors due to noises, the searching process here is designed to search back one step to (4) where the duty cycle is 2.385 μs. When it is confirmed by the controller that the source current at (4) is larger than the source current in (3), the control changes the duty cycle back to 2.26 μs at (5).

The measured efficiencies are compared with the simulation values obtained from the circuit simulator, as shown in Fig. 8.9. The changing trends of the curves are the same and the maximum efficiency point can be found in both the practical system and the simulation.

**Fig. 8.9** Efficiency curves as a function of the pulse width of the output voltage of the inverter with $R_L = 2.5$ Ω: top curve-simulation; bottom curve-measurement

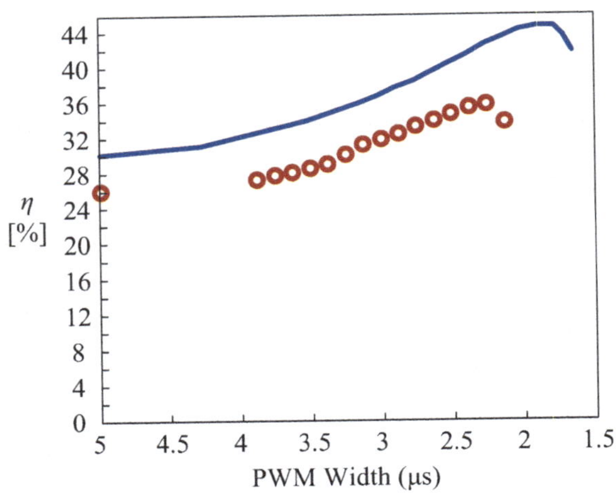

## 8.4 Conclusion

A maximum efficiency tracking method is presented in this chapter with the support of analysis and the experimental verification of a two-resonator WPT system. The basic principle is to search for the minimum input power for any given output power. Another advantage of this method is that it does not need any wireless communications between the transmitter and the receiver circuits, making it attractive in practical applications.

# Chapter 9
# Transmitter-Side On–Off Keying Modulation

As mentioned in Chap. 7, transmitter-side OOK modulation can realize MEO without adding any extra components or changing any hardware of a WPT system which typically consists of an inverter in the primary side and a diode bridge rectifier in the secondary side.

## 9.1 Introduction

Generally speaking, in the design stage of an SS WPT system with a constant output voltage, the optimal load resistance of the system will be preferably designed to equal the minimum load resistance (i.e., the equivalent load resistance value at the rated or maximum power). When the actual load power is less than the rated power, the equivalent load resistance will increase because it is assumed that the WPT system has a constant output voltage. Therefore, only the situation that the load resistance larger than the optimal load (i.e., the lowest equivalent load resistance in the design) is a concern. Therefore the transmitter-side OOK modulation can be used to achieve MEO.

This method is based on an underlying principle that, "for any SS WPT system with a constant output voltage, there will be an optimal input voltage for the system at which a maximum efficiency can be achieved regardless of load changes". So firstly we determine the optimal input voltage of the SS WPT system that is designed to operate at a certain output voltage. Then the OOK modulation ensures that the WPT system operates at the optimal efficiency condition whenever the system is energized.



W. Zhong et al., *Wireless Power Transfer*, CPSS Power Electronics Series,
https://doi.org/10.1007/978-981-15-2441-7_9

## 9.2 Derivation of the Constant-Input-Voltage Principle

### 9.2.1 *Theoretical Analysis on WPT Systems with Output Rectifiers*

Figure 9.1a shows a typical SS WPT system with a given $v_S$ and a full-bridge rectifier. A simplified model, as shown in Fig. 9.1b, is built on the following assumptions:

- The voltage drop of the diode rectifier is constant;
- Only the fundamental components of $v_D$ and $v_O$ are considered;
- The input current of the rectifier is continuous and sinusoidal.

Therefore,

$$R_{EL} = \frac{8}{\pi^2} R_L. \tag{9.1}$$

The circuit equations of the system operating at the resonant frequency of the resonators in Fig. 9.1b are

$$R_{P1}\mathbf{I_1} + j\omega M_{12} I_2 = \mathbf{V_S} \tag{9.2}$$

$$j\omega M_{12}\mathbf{I_1} + (R_{P2} + R_{EL}) I_2 + V_{D1} = \mathbf{0}, \tag{9.3}$$

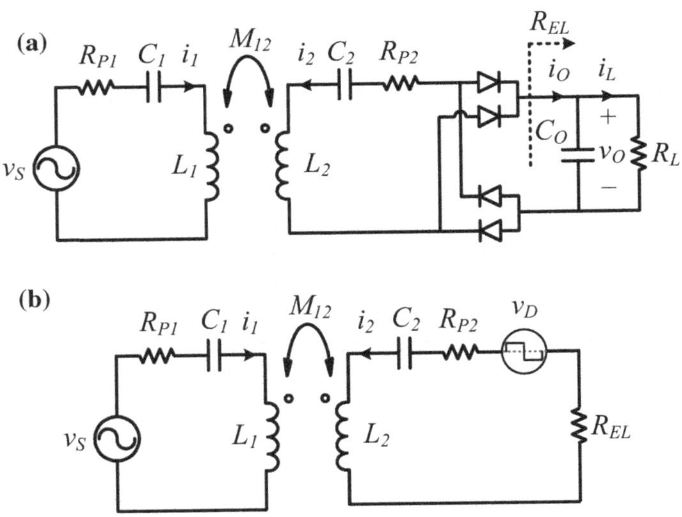


**Fig. 9.1** **a** An SS WPT system with a full-bridge diode rectifier; **b** simplified model of the system

where $V_{D1}$ is the RMS value of the fundamental component of $v_D$. Here $V_{D1}$ and $I_2$, which are in phase, are assumed to have zero phase angle. Then by rearranging (9.3),

$$\mathbf{I_1} = j\frac{(R_{P2} + R_{EL})I_2 + V_{D1}}{\omega M_{12}} \tag{9.4}$$

$$\begin{aligned}\mathbf{V_S} &= R_{P1}\mathbf{I_1} + j\omega M_{12}I_2 \\ &= j\left(\frac{(R_{P2} + R_{EL})I_2 + V_{D1}}{\omega M_{12}}R_{P1} + \omega M_{12}I_2\right).\end{aligned} \tag{9.5}$$

Therefore, $\mathbf{V_S}$ has a 90° phase angle and thus it can be expressed as

$$\mathbf{V_S} = jV_S. \tag{9.6}$$

By solving (9.2), (9.3), and (9.6),

$$\mathbf{I_1} = j\frac{V_S(R_{P2} + R_{EL}) + \omega M_{12}V_{D1}}{\omega^2 M_{12}^2 + R_{P1}(R_{P2} + R_{EL})} \tag{9.7}$$

$$I_2 = \frac{\omega M_{12}V_S - R_{P1}V_{D1}}{\omega^2 M_{12}^2 + R_{P1}(R_{P2} + R_{EL})}. \tag{9.8}$$

The efficiency of the system with switching losses neglected can be expressed as

$$\eta = \frac{I_2^2 R_{EL}}{I_1^2 R_{P1} + I_2^2(R_{P2} + R_{EL}) + V_{D1}I_2}. \tag{9.9}$$

By substituting (9.7) and (9.8) into (9.9) and putting $\frac{\partial \eta}{\partial R_{EL}} = 0$, the optimal load resistance can be obtained as:

$$R_{EL\_OPT} = \frac{1}{V_S}\sqrt{\frac{\Lambda}{R_{P1}}}, \tag{9.10}$$

where

$$\begin{aligned}\Lambda = {} & R_{P1}(\omega M_{12}V_D + R_{P2}V_S)^2 + R_{P2}(\omega M_{12}V_S - R_{P1}V_D)^2 \\ & + \frac{2\sqrt{2}}{\pi}V_D(\omega M_{12}V_S - R_{P1}V_D)\left(\omega^2 M_{12}^2 + R_{P1}R_{P2}\right).\end{aligned}$$

According to (9.10), the optimal load resistance will be dependent on the input voltage when the voltage drop of the diodes is included.

### 9.2.2 Theoretical Analysis on WPT Systems with Constant Output Voltage

Batteries are common loads in WPT applications. The common output of an SS WPT system is a constant voltage source and there will be a rectification stage before the output stage. Therefore, the schematic of the system is redrawn in Fig. 9.2a. Similar assumptions are made to simplify the analysis and the simplified model is shown in Fig. 9.2b. The mathematical expressions are shown below.

$$R_{P1}\mathbf{I_1} + j\omega M_{12} I_2 = \mathbf{V_S} \tag{9.11}$$

$$j\omega M_{12}\mathbf{I_1} + R_{P2} I_2 + V_{D1} + V_{O1} = \mathbf{0}, \tag{9.12}$$

where $V_{D1}$ and $V_{O1}$ are the RMS values of the fundamental components of $v_D$ and $v_O$, respectively. Here $V_{D1}$, $V_{O1}$, and $I_2$, which are in phase, are assumed to have zero phase angle. It should be noted that the simplifications will bring some errors, which will be observed in the experimental results. Nevertheless, the simplified model provides insights into the performance of the system.

Similarly, $\mathbf{V_S}$ has a 90° phase angle and can be represented with (9.6). By solving (9.11) and (9.12),

**Fig. 9.2** Modeling of an SS WPT system with constant output voltage

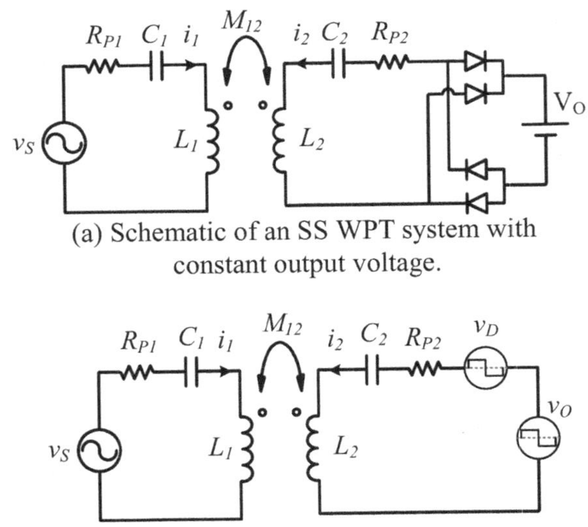


(a) Schematic of an SS WPT system with constant output voltage.

(b) Simplified mode of an SS WPT system with constant output voltage.

$$\mathbf{I}_1 = j\frac{V_S R_{P2} + \omega M_{12}(V_{D1} + V_{O1})}{\omega^2 M_{12}^2 + R_{P1} R_{P2}} \tag{9.13}$$

$$I_2 = \frac{\omega M_{12} V_S - R_{P1}(V_{D1} + V_{O1})}{\omega^2 M_{12}^2 + R_{P1} R_{P2}}. \tag{9.14}$$

The efficiency of the system with switching losses neglected can be expressed as

$$\eta = \frac{V_{O1} I_2}{I_1^2 R_{P1} + I_2^2 R_{P2} + (V_{D1} + V_{O1}) I_2}. \tag{9.15}$$

Then the optimal input voltage can be obtained as

$$V_{S\_OPT} = \frac{-b + \sqrt{b^2 - 4ac}}{2a}, \tag{9.16}$$

where

$$a = \omega M_{12} R_{P1} R_{P2}^2 + \omega^3 M_{12}^3 R_{P2};$$
$$b = -2 R_{P1} R_{P2}\left(\omega^2 M_{12}^2 + R_{P1} R_{P2}\right)(V_{D1} + V_{O1});$$
$$c = -\omega M_{12} R_{P1}\left(\omega^2 M_{12}^2 + R_{P1} R_{P2}\right)(V_{D1} + V_{O1})^2.$$

From the analysis, it can be seen that the optimal input voltage of an SS WPT system can be expressed without considering the load resistance if the output DC voltage is well regulated at a constant level.

## 9.3 An OOK Modulated WPT System

Figure 9.3a shows the schematic of a two-coil SS WPT system based on the OOK modulation. Two frequencies are involved in the switching control of the power inverter as shown in Fig. 9.3b. $S_{HF}$ is the original high-frequency switching signal for the inverter and $S_{LF}$ is the low-frequency OOK modulation signal. When $S_{LF}$ is logic high, the inverter operates normally and when $S_{LF}$ is logic low, the switching actions of the inverter are effectively disabled. In this way, the output power of the system can be regulated by controlling the duty cycle of the low-frequency signal $S_{LF}$. The typical switching action of the transmitter circuit is shown as $v_{in}$ in Fig. 9.3b and the control scheme of the system is given in Fig. 9.3c.

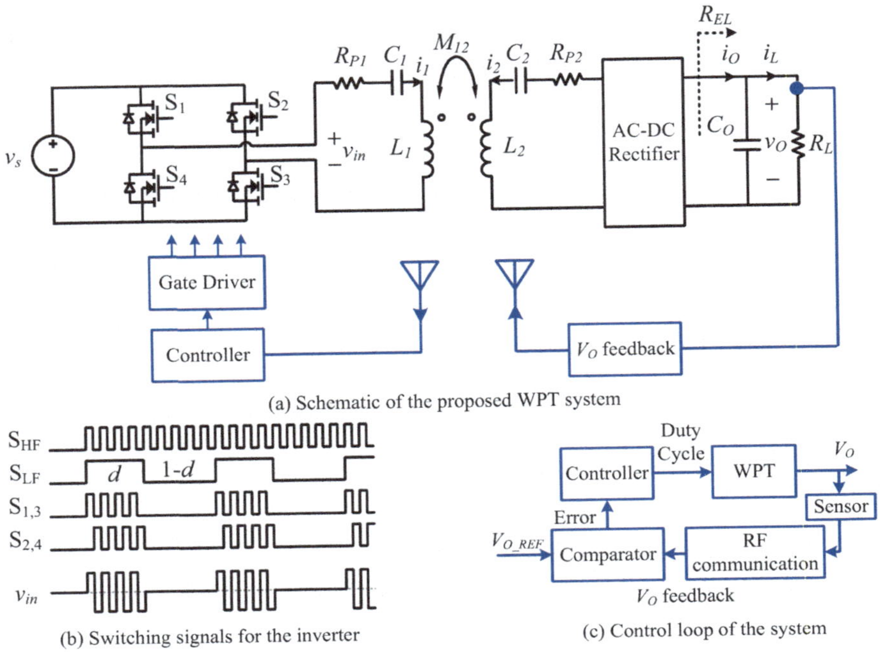


**Fig. 9.3** The OOK modulated SS WPT system

### 9.3.1 Analysis on the Effect of OOK

To simplify the analysis, it is assumed that the transient process only takes up a small part of the whole operating time and thus the only steady state is considered. This assumption should be valid when $S_{HF}/S_{LF}$ is large enough. When the transient process is neglected, the efficiency of the OOK modulated system will be equal to the efficiency of the system being activated, i.e., when $S_{LF}$ is logic high. It is important to note that the when the system is activated, the currents in the windings are given by (9.13) and (9.14), which implies that the operation state of the system will keep the same (i.e., the same input and output voltages, and the same input and output currents) regardless of the load power. This is the reason why the efficiency of the OOK modulated system is independent of the load as indicated by (9.16).

The actual load power may vary with time. To match the rectifier power output and the actual load power requirements, the average output power could be regulated by adjusting the duty cycle of the modulation signal as:

$$P_L = dP_O, \tag{9.17}$$

where $P_O$ is the output power of the output rectifier when the system is activated, and $d$ is the duty cycle of $S_{LF}$.

Now the duty cycle of $S_{LF}$ is $d$ ($0 < d < 1$). The current injected into the output capacitor and the load is $i_O$ when $S_{LF}$ is high and the load current is $i_L$ as marked in Fig. 9.3. Therefore, the output energy of the rectifier should be equal to the energy consumption of the load, i.e.,

$$V_{O1} I_O d T = V_O I_L T, \tag{9.18}$$

where $T$ is the period of $S_{LF}$, $V_O$ is DC output voltage, and $V_{O1}$ is the RMS value of the fundamental component of $v_D$ as defined above. Therefore,

$$V_{O1} = \frac{2\sqrt{2}}{\pi} V_O \tag{9.19}$$

$$I_O = \frac{\pi \cdot I_L}{2\sqrt{2} d}. \tag{9.20}$$

Define $R_{EL}$ (see Fig. 9.3) as the resistance through which $i_O$ will generate the same amount of power as that injected into the capacitor and the load, i.e., $V_{O1} I_O$. It should be noted that the efficiency of the system calculated with $R_{EL}$ is not exactly equal to the practical value because $R_{EL}$ is an equivalent value from the aspect of power. However, it is close enough for practical purpose.

$$I_O^2 R_{EL} = V_{O1} I_O \tag{9.21}$$

Therefore, by substituting (9.20) into (9.21), the equivalent load resistance can be obtained as

$$R_{EL} = \frac{V_{O1}}{I_O} = d \frac{8}{\pi^2} \frac{V_O}{I_L} = d \frac{8}{\pi^2} R_L. \tag{9.22}$$

Equation (9.22) shows that the equivalent load resistance $R_{EL}$ can be adjusted by changing the duty cycle $d$ of $S_{LF}$. In practice, the maximum rated output power corresponds to the minimum value of $R_{L\text{-min}}$, at which the duty cycle ($d$) should become 1. When the actual load resistance value is larger than $R_{L\text{-min}}$, the duty cycle will be smaller than 1. The control of the duty cycle is the basis of this OOK method. Whenever the WPT system is activated (i.e., ON time of the control), power transfer occurs at the maximum-efficiency condition which is ensured by matching the optimal load resistance value as given in (9.10) to the actual $R_{L\text{-min}}$ in the design stage. This makes the control of the SS WPT system very simple and effective in achieving high efficiency over the load range.

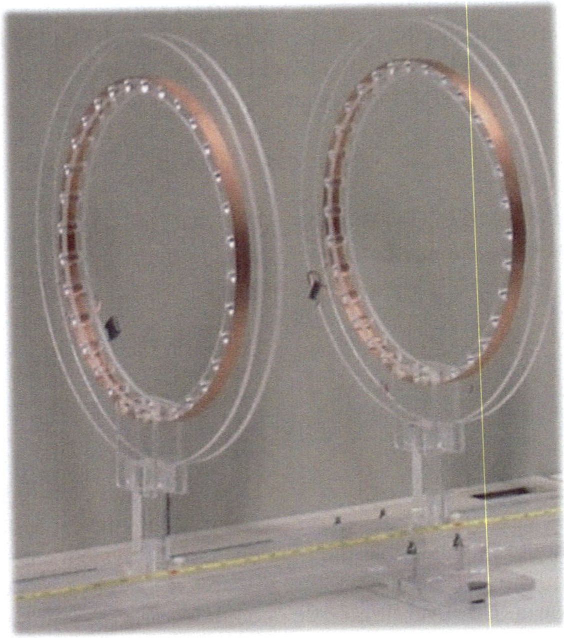

**Fig. 9.4** Coil-resonators used for simulation study and experiments

### *9.3.2 Simulation Study*

To illustrate the design process for such a control scheme, we use the parameters of practical coils (Fig. 9.4) for the following simulation study. The parameters are listed in Table 9.1.

According to (9.16), the optimal input voltage of the system is 11.2 V for an output voltage of 10 V. The theoretical optimal duty cycle of the OOK modulation signal can be expressed as a function of the load resistance, as shown in Fig. 9.5. The variation of the output power of the system with respect to the duty cycle is shown in Fig. 9.6. The maximum output power of the system is 23.2 W.

## 9.4 Experimental Verifications

Experiments have been carried out based on the two-coil system. Figure 9.7 shows the inverter and the microprocessor in the primary side. MOSFET DMT6016LSS is used in the full-bridge inverter. The microprocessor is STM32F100RBT6B. Schottky diode PMEG3050EP is used in the full-bridge output rectifier.

Practical measurement results are recorded in Figs. 9.8 and 9.9. Figure 9.8 shows the lower frequency switching signal $\overline{S}_{LF}$ (i.e., the inverted $S_{LF}$, upper trace), the input voltage of the primary resonator $v_{in}$ (middle trace), and the output voltage of

**Table 9.1** Parameters of a WPT system

| | |
|---|---|
| Resonant frequency $f_0$ | 97.56 kHz |
| Radius of windings | 155 mm |
| Number of turns | 11 |
| Layers of the wire | 1 |
| Structure of the wire | Ø0.12 mm × 50 strands<br>Outer Ø1.2 mm |
| $L_1$ | 91.24 μH |
| $L_2$ | 91.77 μH |
| Distance of the windings, $d_{12}$ | 200 mm ($k = 0.0689$) |
| $R_{P1}$ | 0.441 Ω (including the on-state resistance of the MOSFETs in the inverter) |
| $R_{P2}$ | 0.415 Ω |
| $R_L$ | 20 Ω |
| $V_O$ | 10 V |
| $V_D$ | 2 × 0.4 V |

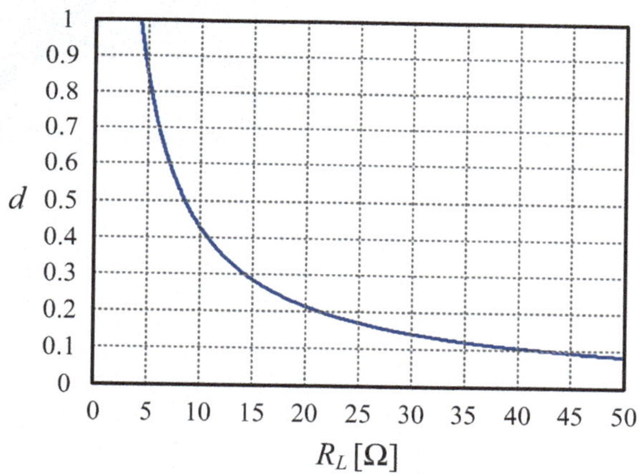


**Fig. 9.5** Variation of the duty cycle of the OOK modulation signal with the load resistance

the system $v_o$ (lower trace). The frequency of $\overline{S}_{LF}$ is 1 kHz. Figure 9.9 shows the input voltage of the primary resonator and the output voltage when the system is activated. The high-frequency switching signal in the middle trace is 97.56 kHz.

Figure 9.10 shows the system efficiency variations as the input voltage changes under three loading conditions which are approximately

- 10 Ω load resistance and 10 W output power (43% rated load);
- 20 Ω load resistance and 5 W output power (22% rated load);
- 40 Ω load resistance and 2.5 W output power (11% rated load).

The observations from Fig. 9.10 and the relevant discussions are listed below.

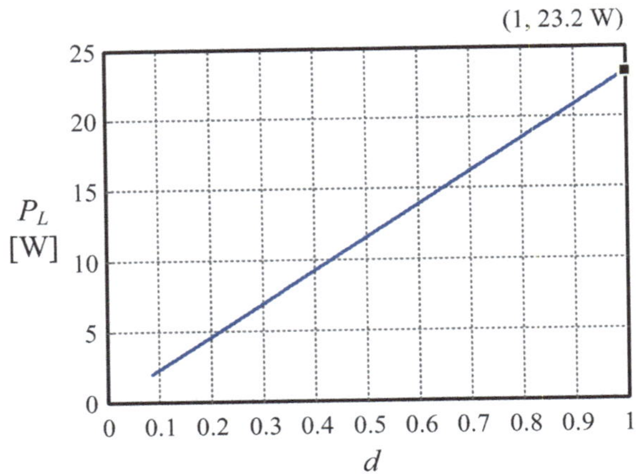


**Fig. 9.6** Variation of the load power with the duty cycle of the OOK modulation signal

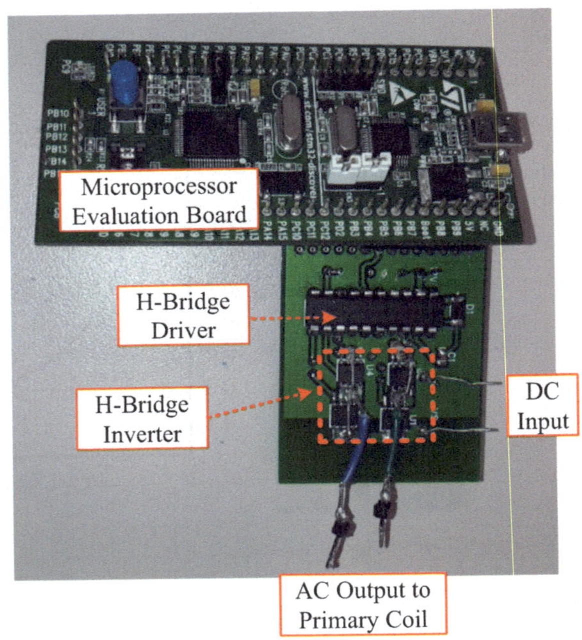


**Fig. 9.7** Primary circuit of the OOK modulated WPT system

(1) All three measured curves peak at the same input voltage range.
(2) The system efficiencies are improved significantly by changing the input voltage for all three loads. Table 9.2 provides a comparison between several operation scenarios.

- **Theoretical System Efficiency without Transforming Load Resistance**: without transforming load resistance, the efficiencies are the lowest.

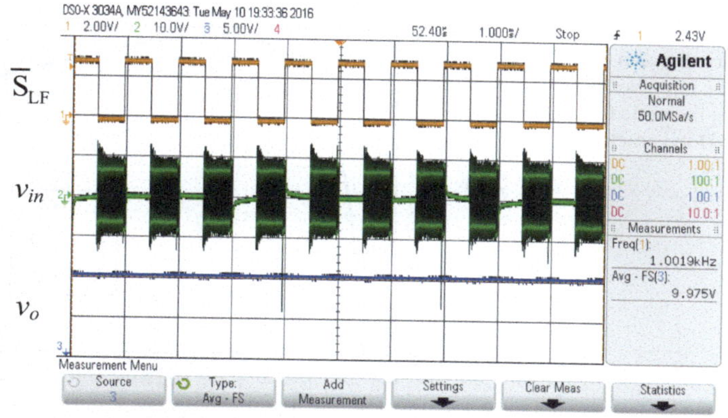


**Fig. 9.8** Top waveform: Inverted signal of $S_{LF}$ (2 V/div.); Middle waveform: voltage applied on the primary resonator $v_{in}$ (10 V/div.); Bottom waveform: output voltage of the system $v_o$ (5 V/div.). Time division is 1 ms/div

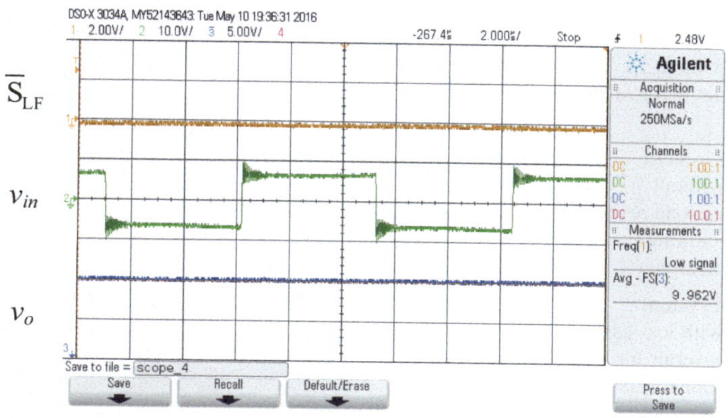


**Fig. 9.9** Waveforms of the input and output voltages when the system is activated. Top waveform: Inverted signal of $S_{LF}$ (2 V/div.); Middle waveform: voltage applied on the primary resonator $v_{in}$ (10 V/div.); Bottom waveform: output voltage of the system $v_o$ (5 V/div.). Time division is 2 μs/div

- **Theoretical System Efficiency without Transforming Load Resistance and NOT Counting the loss of the rectification**: For 20 and 40 Ω cases, the efficiencies of the OOK system which have included the rectification losses are even higher than those in this scenario, which implies that the OOK system can transform the load resistance to the optimal value.
- **Theoretical System Efficiency with Lossless DC–DC Converter for Load Resistance Transformation**: In other words, the actual load resistance has been perfectly matched to the optimal value without causing any losses. The efficiencies of the OOK system are even higher than those in this scenario, which implies that the proposed system is superior over the systems using

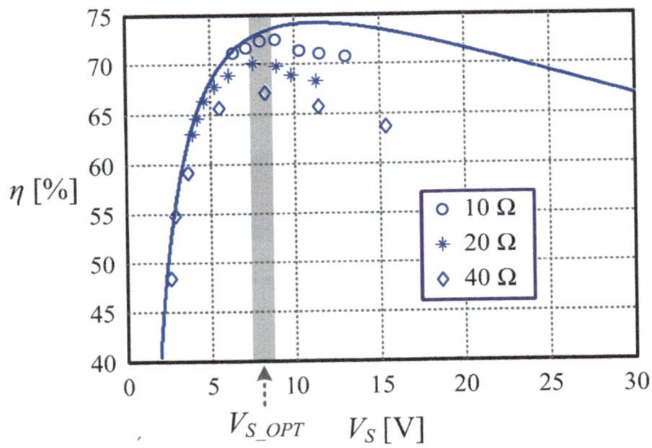


**Fig. 9.10** Overall system efficiency variation as a function of the input voltage $V_S$. Real line: theoretical calculations; Dots: practical measurements

**Table 9.2** Comparison among several operation scenarios

| | 10 Ω (43% $P_{Lrated}$) | 20 Ω (22% $P_{Lrated}$) | 40 Ω (11% $P_{Lrated}$) |
|---|---|---|---|
| Theoretical system efficiency without transforming load resistance (%) | 69.58 | 59.15 | 44.73 |
| Theoretical system efficiency without transforming load resistance and NOT counting the loss of the rectification (%) | 76.01 | 65.4 | 50.13 |
| Theoretical system efficiency with lossless DC–DC converter for load resistance transformation (%) | 71.90 | 69.01 | 65.32 |
| Measured efficiency of the OOK system (%) | 72.60 | 70.20 | 67.27 |

DC–DC converters for load resistance transformation. The reason is that the OOK system can not only convert the actual load resistance to the optimal value, but also reduce the loss of the rectification stage at the receiver side.

(3) When the input voltage is high, the errors between the measurements and the theoretical values become larger. There are two major reasons. Firstly, three assumptions are made in the simplified model. Secondly, there is a transient process when the OOK method engages the transmitter circuit for WPT with the receiver circuit as shown in Fig. 9.8. The extra power losses in these transients are neglected in the analysis. Therefore, when the input voltage is higher, the transient processes will be more obvious and affect the results more. Also

because of these reasons, the practical optimal input voltage is not exactly the same as the theoretical one.

## 9.5 Conclusion

This chapter introduces the OOK modulation to maintain high-efficiency operation for SS WPT systems over a wide load range. By determining the optimal input voltage for a given output voltage in an SS WPT system, MEO can theoretically be achieved independently of the load. This simple method can be implemented in SS WPT systems without adding extra power electronics.

# Chapter 10
# Reconfigurable WPT Systems—A Design Example

## 10.1 Introduction

In this chapter, a reconfigurable structure that combines the functions of a matching network and the transformer mechanism which is used in the three-resonator system is investigated. No additional coil is used because the different portions of the same receiving coil are used to realize the transformer action. The idea is to design the optimal efficiency ranges of two or more resonant tanks' characteristics so that their combined high-efficiency range covers a very wide range of the load. Near maximum efficiency can be achieved for a load resistance in the range roughly from 1/10 to 20 times of the optimal load resistance of the original two-resonator system.

## 10.2 Use Receiving Coil Splitting to Enable High Efficiency for Smaller Load Resistances

As shown in Fig. 10.1, the receiving coil, i.e., Coil-S, is split into two coils Coil-S1 and Coil-S2 while the transmitting coil, i.e., Coil-P, is unchanged. Figure 10.2 shows the circuit model of the resultant three-coil WPT system where $R_P$ is the resistance in the primary loop, which normally includes the source resistance of the power source and the parasitic resistance of the coil and the compensation capacitor; $R_{S1}$ and $R_{S2}$ are the parasitic resistance in respective current loops; $R_L$ is the load resistance; $L_P$, $L_{S1}$, $L_{S2}$, $C_P$, $C_{S1}$, $C_{S2}$ are the respective self-inductance of the coils and the compensating capacitance; $M_{PS1}$ and $M_{PS2}$ are the mutual inductance between the primary coil and the first splitting coil and the second splitting coil, respectively; $M_{12}$ is the mutual inductance between two splitting coils.

If the loaded coil is Coil-S2 in the three-coil system, there are two possible paths for power transfer, i.e., "Coil-P to Coil-S2" and "Coil-P to Coil-S1 to Coil-S2". Assume that Coil-S1 has most of the turns of the original coil and Coil-S2 has a



W. Zhong et al., *Wireless Power Transfer*, CPSS Power Electronics Series,
https://doi.org/10.1007/978-981-15-2441-7_10

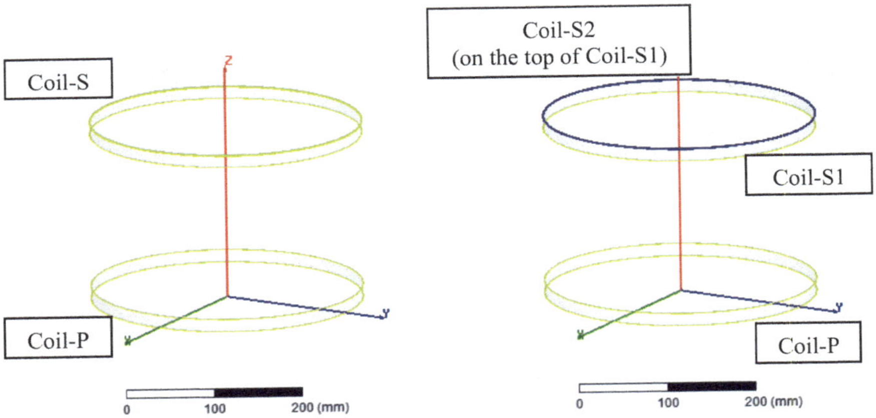


**Fig. 10.1** Receiving coil is split into two coils

**Fig. 10.2** Schematic of the system with receiving coil split into two

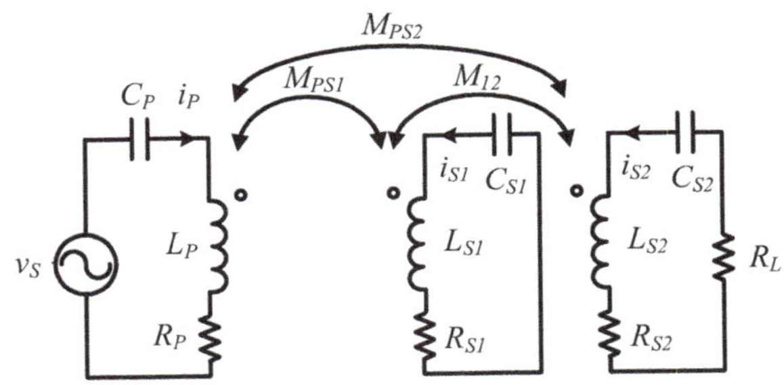


small portion of the turns of the original coil, e.g., 1 turn. Then the first power path can be neglected because the mutual inductance $M_{PS2}$ will be much smaller than the mutual inductance $M_{PS1}$ and $M_{12}$. So most of the power will be delivered to Coil-S2 through the second power path in which Coil-S1 performs as a relay coil. Also, the power loss in Coil-S2 is negligibly small because the resistance of Coil-S2 is very small. Assuming the coils are resonant at the operating frequency, the simplified circuit equations of the system without considering the first power path and the power loss in Coil-S2 can be expressed as follows.

$$R_P\mathbf{I_P} + j\omega M_{PS1}\mathbf{I_{S1}} = V_S \tag{10.1}$$

$$j\omega M_{PS1}\mathbf{I_P} + R_{S1}\mathbf{I_{S1}} + j\omega M_{12}\mathbf{I_{S2}} = \mathbf{0} \tag{10.2}$$

$$j\omega M_{12}\mathbf{I_{S1}} + R_L\mathbf{I_{S2}} = \mathbf{0}. \tag{10.3}$$

The reflected load resistance in Coil-S1 is

$$R_{L\_S1} = \frac{\omega^2 M_{12}^2}{R_L}. \tag{10.4}$$

Setting

$$R_{L\_S1} = R_{L\_OPT}, \tag{10.5}$$

where $R_{L\_OPT}$ is the optimal load resistance of the original two-resonator system as defined by. Then,

$$M_{12} = \frac{\sqrt{R_L R_{L\_OPT}}}{\omega}. \tag{10.6}$$

Therefore, the smaller the $R_L$ is, the smaller the required $M_{12}$ should be in order to match the actual load resistance to the optimal value.

A case study is provided to prove that the three-resonator system with the splitting receiving coil will have a smaller optimal load resistance. The study is based on two identical coils as shown in Fig. 10.3 with parameters listed in Table 10.1. In this study, the receiving coil is split into two coils to form two selectable coil-resonators with resonance capacitors, as shown in Fig. 10.4. The upper split coil has $n$ turns ($n = 1, 2, \ldots, 10$) and thus the lower one has $(11 - n)$ turns. The operating frequency is set at 100 kHz. At this frequency, the AC resistance of the coils is

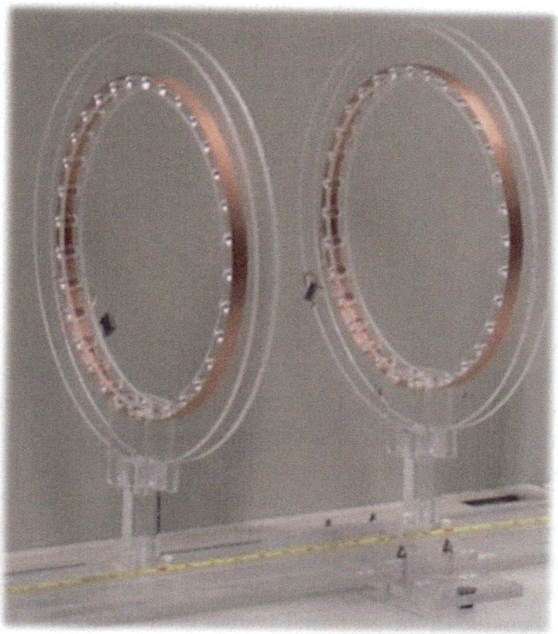

**Fig. 10.3** Coils used for simulation study and experiments

**Table 10.1** Parameters of the WPT system used for simulations and experiments

| | |
|---|---|
| Radius of windings | 155 mm |
| Number of turns | 11 |
| Layers of the wire | 1 |
| Structure of the wire | Ø0.12 mm × 50 strands<br>Outer Ø1.2 mm |
| Self-inductance | 91.25 μH |
| Parasitic resistance | 0.394 Ω ($Q$ = 146; 100 kHz) |
| Distance between the transmitting and receiving coils | 100 mm ($k$ = 0.179) |

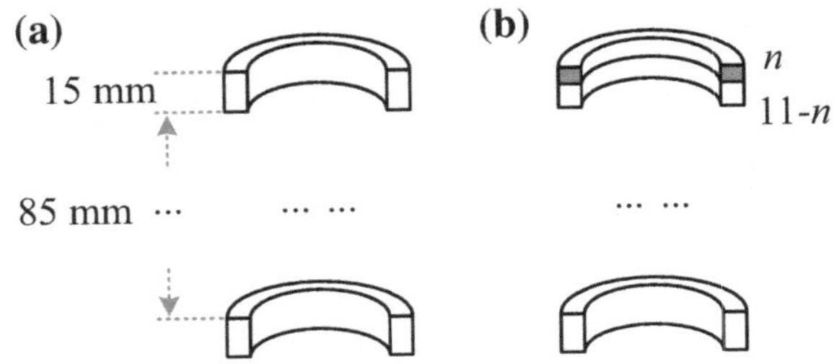


**Fig. 10.4** **a** Original two-resonator system with the bottom coil as the transmitting coil; **b** receiving coil split into two coils

negligibly small as reflected in the practical measurements (about 6% of the DC resistance). Therefore, a simplification is made that the resistance of the split coils is proportional to the number of turns. The upper coil is connected to the load.

The general equations for the three-resonator system shown in Fig. 10.4 are

$$(R_P + jX_P)\mathbf{I_P} + j\omega M_{PS1}\mathbf{I_{S1}} + j\omega M_{PS2}\mathbf{I_{S2}} = V_S \tag{10.7}$$

$$j\omega M_{PS1}\mathbf{I_P} + (R_{S1} + jX_{S1})\mathbf{I_{S1}} + j\omega M_{12}\mathbf{I_{S2}} = \mathbf{0} \tag{10.8}$$

$$j\omega M_{PS2}\mathbf{I_P} + j\omega M_{12}\mathbf{I_{S1}} + (R_{S2} + R_L + jX_{S2})\mathbf{I_{S2}} = \mathbf{0}, \tag{10.9}$$

where $X_{P,S1,\text{or }S2} = \omega L_{P,S1,\text{or }S2} - \frac{1}{\omega C_{P,S1,\text{or }S2}}$. In order to maximize the efficiency of the given system, $C_{S1}$, $C_{S2}$, and $R_L$ can be optimized.

The number of turns of the upper split coil (connected to the load) is changed from 1 to 10 in a step-by-step manner as shown in Fig. 10.4. The optimization results are shown and labeled as Case 1 in Figs. 10.5, 10.6 and 10.7.

Case 2 is the optimization under the constraint that $C_{S1}$ in series of $C_{S2}$ will meet the required capacitance for the whole receiving coil to resonate at the operating frequency (100 kHz). An important feature of this research idea is to enable the WPT system flexibly to operate in either two-coil mode or three-coil mode. The schematic of the proposed system is shown in Fig. 10.8 and will be analyzed in the

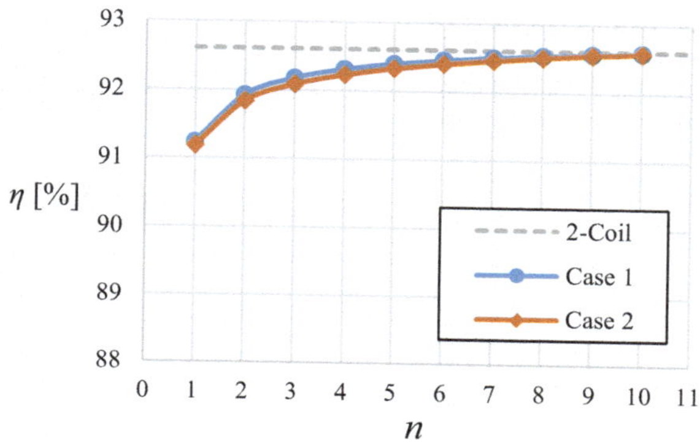


**Fig. 10.5** Maximum efficiency versus $n$ for the two-resonator, Case 1 and Case 2 configurations

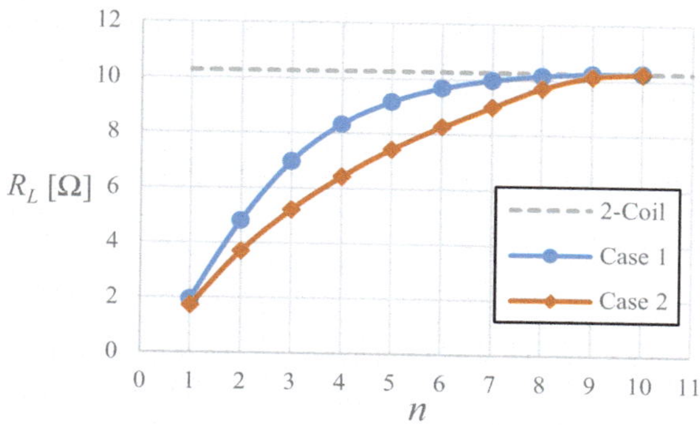


**Fig. 10.6** Optimal load resistance versus $n$ for the two-resonator, Case 1 and Case 2 configurations

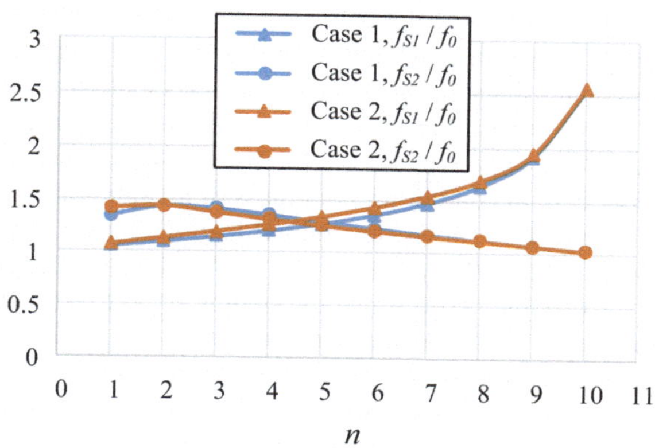


**Fig. 10.7** Individual resonant frequency of the split coil-resonators normalized at the operating frequency of 100 kHz

next section. Therefore, the conditions of the optimization of Case 1 and Case 2 can be expressed mathematically as:

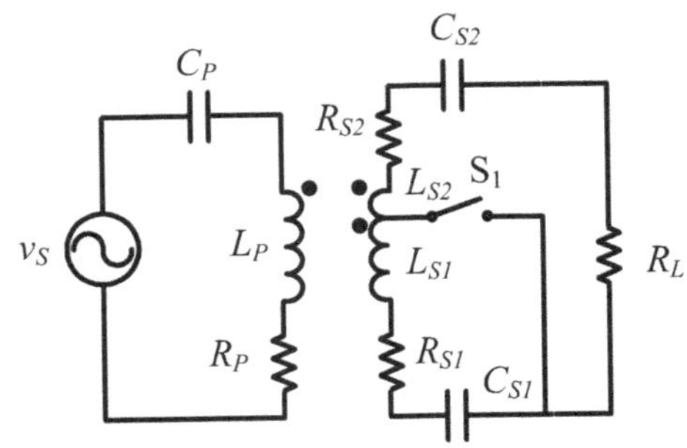


**Fig. 10.8** A proposed WPT system which can operate in either the three-resonator mode or the two-resonator mode

Case 1:

$$\max: \quad \eta = \mathrm{f}(C_{S1}, C_{S2}, R_L) \tag{10.10}$$

Case 2:

$$\max: \quad \eta = \mathrm{f}(C_{S1}, C_{S2}, R_L)$$

$$\text{with constraint:} \quad \frac{1}{\omega C_{S1}} + \frac{1}{\omega C_{S2}} = \omega L_S, \tag{10.11}$$

where $L_S$ is the self-inductance of the whole secondary coil. For each $n$, a set of optimal values for $C_{S1}$, $C_{S2}$, and $R_L$ will be determined.

Figure 10.5 shows the maximum efficiency curves versus $n$ under the two cases. It is noted that the efficiency curves of the two cases of coil splitting are very similar. When $n$ is small, there is only a small efficiency drop (of about 1.4% when $n = 1$) when compared with the maximum efficiency of the original two-resonator structure. This observation on splitting receiving coil is consistent with the previous result presented in Chap. 7 that a three-resonator system with a splitting transmitting coil could have nearly the same maximum efficiency as that of the original two-resonator system under certain conditions, regardless of the ratio of the coil splitting.

Figure 10.6 shows the optimal load resistance value versus $n$ for the two cases. The minimum $R_{L\_\mathrm{OPT}}$ becomes 1.7 Ω when $n = 1$ while the optimal load resistance of the original two-resonator system is 10.3 Ω. When $n$ becomes larger, the optimal load value of the three-resonator system approaches the optimal load value of the two-resonator system. Therefore, splitting the receiving coil does have the ability to shift the optimal load resistance to a smaller value. This interesting feature provides a useful mechanism to achieve high efficiency over a wide load range.

Figure 10.7 shows the variations of the per-unit ratios of the resonance frequencies of the split receiving coils with the optimized compensating capacitance ($f_{S1}$ for the upper split coil-resonator and $f_{S2}$ for the lower split coil-resonator) using the operating frequency of the system $f_0$ (i.e., 100 kHz) as the base value. It can be observed that the two curves of $f_{S1}/f_0$ almost overlap for Case 1 and Case 2. Similarly, the two curves of $f_{S2}/f_0$ also overlap for the two cases. These two sets of curves indicate that the resonance frequencies of the splitting receiving coils with different turns ratios do not equal the operating frequency of the system except at the two extreme points (i.e., when $n = 1$ or 10). This feature is different from the operation principle of a two-resonator system. The reason for this difference is that splitting the receiving coil-resonator into two coil-resonators forms a higher order resonant system. Therefore, the individual resonant frequency is not going to dominate the power transfer efficiency of the system.

In this study, the upper split coil is connected to the load. In principle, the lower split coil can be connected to the load if desired. Calculations have been done on these two alternative load arrangements and the results are basically the same. For example, if the load is connected to the upper coil and $n = 1$, the optimal load resistance, capacitance $C_{S1}$, $C_{S2}$ in Case 1, and the maximum efficiency are, respectively, 1.95 Ω, 29.45 nF, 1.255 μF, and 91.23%. If the load is connected to the lower coil and $n = 10$ (loaded coil has one turn), the optimal load resistance, capacitance $C_{S1}$, $C_{S2}$ in Case 1, and the maximum efficiency are, respectively, 1.88 Ω, 28.95 nF, 1.206 μF, and 91.32%.

## 10.3 New Reconfigurable Topologies for Maximizing Efficiency and Power over Wide Load Range

### *10.3.1 Extending High-Efficiency Region to the Lower Load Resistance Range*

The analysis presented in the last section shows that splitting the receiving coil can achieve a maximum efficiency similar to that of the original two-resonator system, but the optimal load resistance value can be much lower. However, either the two-resonator or the three-resonator structure cannot maintain high efficiency within a wide load resistance range. Therefore, a reconfigurable receiver circuit structure as shown in Fig. 10.8 is proposed to enable the WPT system to operate either in the two-resonator mode or the three-resonator mode in order to widen the high-efficiency operation range.

In Fig. 10.8, the receiving coil is dual tapped and a switch is added to switch the secondary circuit to operate in the two-resonator mode or the three-resonator mode. When the switch $S_1$ is turned on, a three-resonator system is formed and the load is connected to $L_{S2}$ (the coil with one turn in this design). Under this condition, $L_{S1}$ (the large coil) and $C_{S1}$ form a coil-resonator and act as a relay or repeater. Thus, the system will operate in the three-resonator mode. When the switch $S_1$ is turned off, the receiver circuit is reverted to a two-resonator system. The whole receiving coil will resonate with the series-connected resonant capacitors $C_{S1}$ and $C_{S2}$. For battery charging applications, low-speed low-cost relay or bidirectional switch could be used for mode switching.

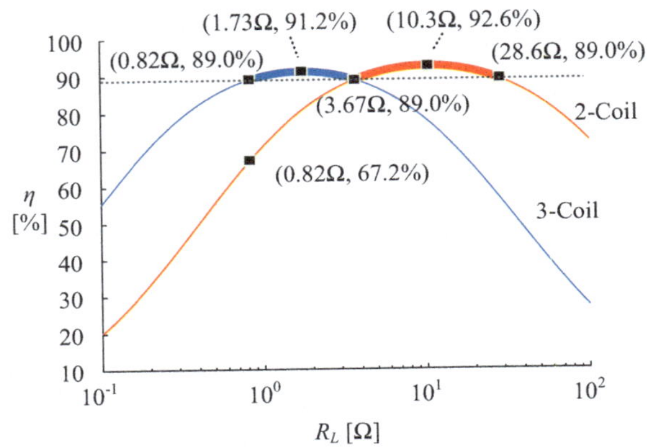


**Fig. 10.9** Efficiency curves of the three-resonator and two-resonator modes ($k = 0.179$)

Figure 10.9 shows the efficiency curves of the two-resonator and the three-resonator configurations of the proposed system as a function of the load resistance. It should be noted that the x-axis uses a logarithmic scale for the load resistance value. The top parts of the two efficiency curves are highlighted in blue color for the three-resonator mode and red color for the two-resonator mode. The mode switching will be conducted at the crossover load resistance point of 3.67 Ω.

Assuming that the design target is to achieve efficiency higher than 89%, the three-resonator mode will extend the lower boundary of the load resistance from 3.67 to 0.82 Ω. Alternatively speaking, if the application needs to operate with 0.82 Ω, the two-resonator system can only achieve efficiency of 67.2% and for the three-resonator system, it is 89%. The improvement is 21.8%.

It should be noted that this case study is a general analysis. It shows that the three-resonator mode is able to extend the high-efficiency load resistance range to much lower load resistance values.

### 10.3.2 Extending High-Efficiency Region to the Higher Load Resistance Range

Furthermore, if the practical load resistance is much larger than the optimal load resistance of the two-resonator system, high efficiency can be achieved by paralleling a capacitor with the load resistance. A second reconfigurable receiver circuit is hereby proposed here and shown in Fig. 10.10. One additional capacitor $C_{LP}$ is connected in parallel with the load and another capacitor $C_{S1P}$ is added to compensate the extra reactance caused by $C_{LP}$. Two more low-speed switches are added for mode switching. Now the system has three operation modes which are

1. Three-resonator mode ($S_1$ on; $S_2$, $S_3$ off)
2. Two-resonator mode ($S_1$ off; $S_2$, $S_3$ off)
3. Two-resonator with parallel capacitor mode ($S_1$ off; $S_2$, $S_3$ on)

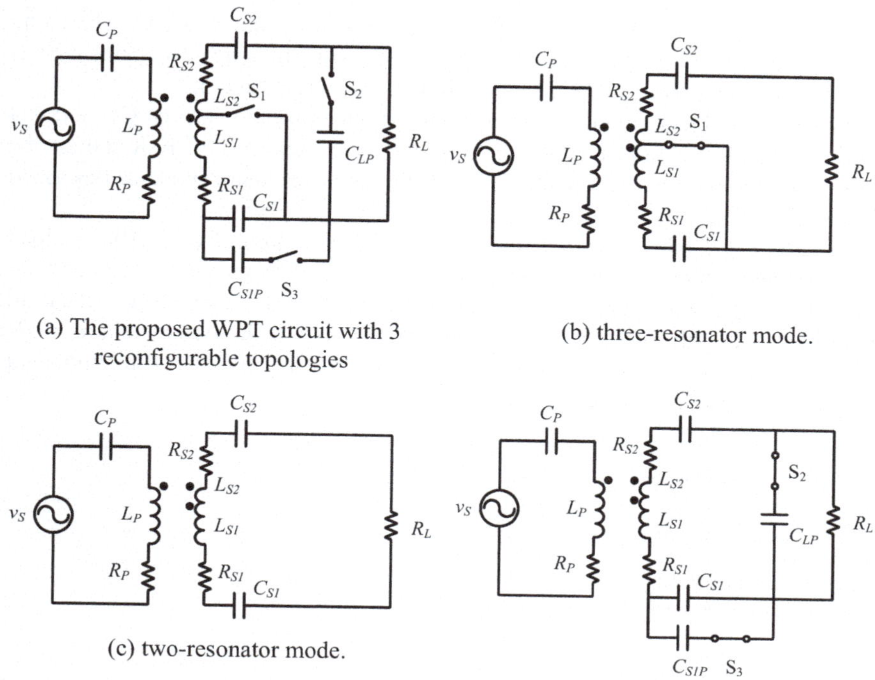


(a) The proposed WPT circuit with 3 reconfigurable topologies

(b) three-resonator mode.

(c) two-resonator mode.

(d) two-resonator with parallel capacitor mode.

**Fig. 10.10** The proposed WPT system with three reconfigurable topologies (**a**) and the equivalent topologies under the three operating modes (**b**), (**c**), and (**d**)

The efficiency curves under the three operating modes of this reconfigurable receiver circuit as a function of the load resistance are shown in Fig. 10.11. The load resistance range with efficiency exceeding 89% for the original two-resonator

**Fig. 10.11** Efficiency curves of the three-resonator, two-resonator, and two-resonator with parallel capacitor modes ($k = 0.179$)

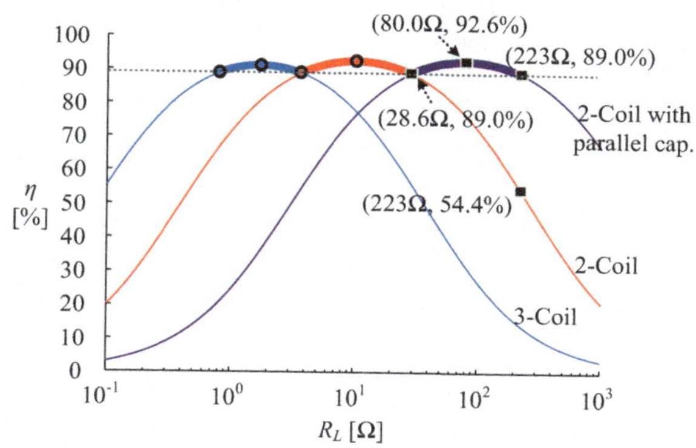

system is from 3.67 to 28.6 Ω (i.e., **7.8** times the minimum load resistance). By using the proposed circuit, the new load resistance range with efficiency exceeding 89% is from 0.82 to 223 Ω (i.e., **272** times the minimum load resistance).

For the second reconfigurable receiver circuit, the optimal values of $C_{S1}$ and $C_{S2}$ are the same 28.5 nF and 1.13 μF as those for the system in Fig. 10.8 and can be worked out by maximizing the efficiency of the three-resonator mode as described in (10.11).

The efficiency of the original two-resonator mode maximizes at 10.3 Ω load resistance and achieves >89% in the load resistance range between 3.67 and 28.6 Ω, as shown in Fig. 10.9. Therefore, one may expect that the efficiency curve of the third operating mode will intersect with the efficiency curve of the two-resonator mode at 28.6 Ω. This means the efficiency of the third operating mode will maximize at about $28.6 \times 28.6/10.3 \approx 80\ \Omega$.

**Fig. 10.12** Required input voltage for 10 W output (blue: three-resonator mode; red: two-resonator mode; purple: two-resonator with parallel capacitor mode)

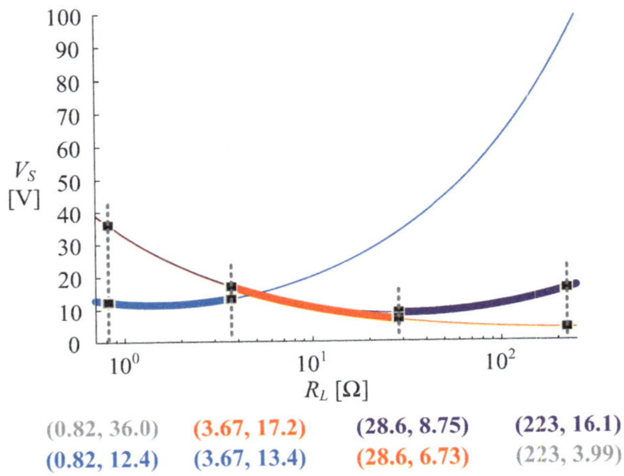


**Fig. 10.13** Required primary coil current for 10 W output (blue: three-resonator mode; red: two-resonator mode; purple: two-resonator with parallel capacitor mode)

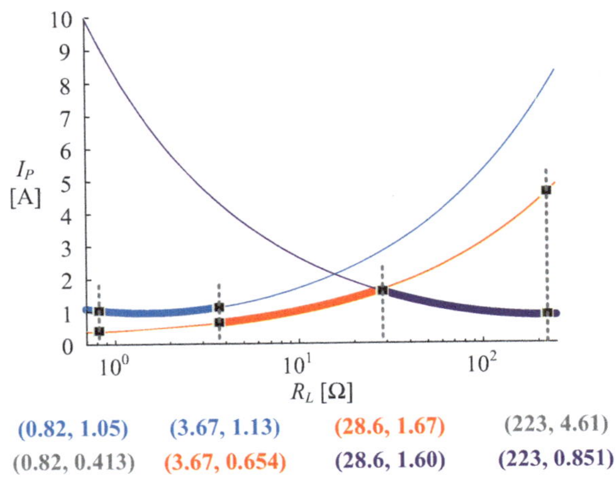

The value of $C_{LP}$ (which can transfer 80 Ω to the optimal load resistance) could be obtained as 51.8 nF. Then the whole secondary circuit in the third operating mode should be resonant at the operating frequency, i.e., the imaginary part of the impedance should be equal to zero which can be expressed as

$$\omega L_2 - \frac{1}{\omega C_{S2}} - \frac{1}{\omega(C_{S1} + C_{S1P})} + \frac{X_{LP} R_L^2}{X_{LP}^2 + R_L^2} = 0. \qquad (10.12)$$

By solving (10.12), $C_{S1P}$ can be obtained as 26.1 nF.

### 10.3.3 VA Rating Minimization or Power Maximization

Apart from the efficiency improvement, the required VA rating of the system can also be minimized by using the proposed reconfigurable WPT system for a given output power, when compared with the original two-resonator system. Here the VA rating means the product of the required input voltage and the primary coil current for a given output power. Figure 10.12 shows the required input voltage for 10 W output. The bolded curves are the operating region for the reconfigurable WPT system. It can be seen that, by switching the operating mode to stay in the region with the low input voltage requirements, this WPT system can be operated in a range of 6.7–16 V for the whole load resistance range of interest. This advantageous feature will not only minimize the voltage rating for the whole system, but also avoid the primary DC–DC converter (feeding the inverter) or the inverter to operate with an extremely small duty cycle.

The current rating is another critical design parameter because it determines the size of the wire used and accordingly the copper cost. Figures 10.13 and 10.14

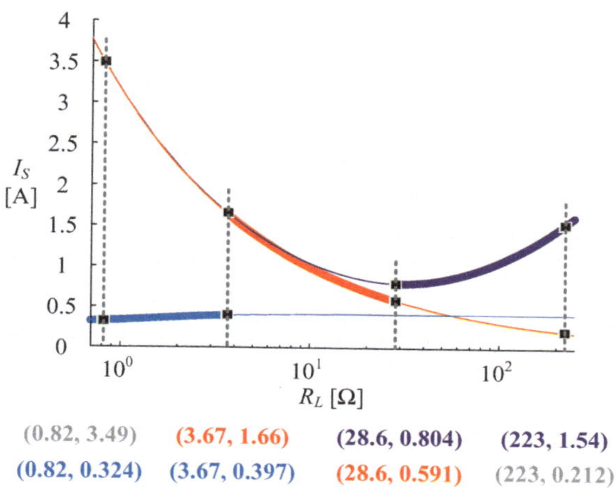


**Fig. 10.14** Required secondary coil current for 10 W output (blue: three-resonator mode; red: two-resonator mode; purple: two-resonator with parallel capacitor mode)

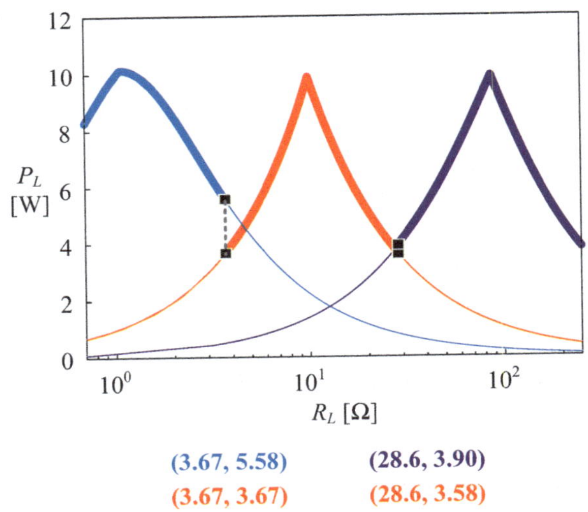


**Fig. 10.15** Maximum output power with the constraint of maximum coil current equals to 1 A output (blue: three-resonator mode; red: two-resonator mode; purple: two-resonator with parallel capacitor mode)

show the required current in the primary coil and secondary coil for 10 W output, respectively. For the two-resonator with parallel capacitor mode, the secondary coil current means the current in the larger coil which takes up most of the copper in the secondary side. The current rating for both the primary and secondary coil of the universal system is about 1.67 A.

Figures 10.12, 10.13, and 10.14 show not only the low VA rating requirements of the system but also the points of mode switching. Mode switching can thus be easily achieved by monitoring the equivalent load resistance. Minimizing the VA rating of the system for a given output power is equivalent to maximizing the output power within given voltage and/or current constraint. Figure 10.15 shows the output power of the reconfigurable WPT system with 1 A limitation for the coil current. It is clear that this WPT system enables the use of the respective maximum power region of all three operating modes.

## 10.4 Experimental Verification

Experiments are carried out based on the 10 W WPT system using the coils shown in Fig. 10.3 and the topology as shown in Fig. 10.10. The receiver coil is split into two coils with 10 turns and 1 turn, respectively, as shown in Fig. 10.16. As mode switching will only be activated after the power supply is shut down, the switches could be easily controlled with microprocessors monitoring the load resistance range. In experiments, mode switching is done manually to measure the efficiency of the system. Table 10.2 shows the parameters of the proposed structure and circuit. Since the lead wires, which are used to connect the coil and the compensation circuit, are not considered in the theoretical analysis, the self-inductances the one-turn coils a

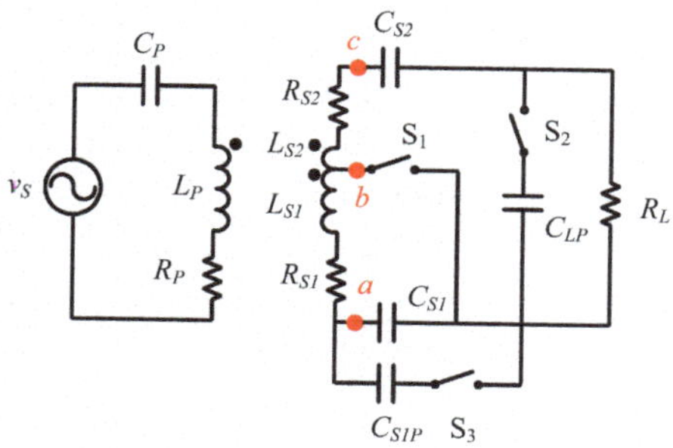


(a) Schematic of the proposed system.

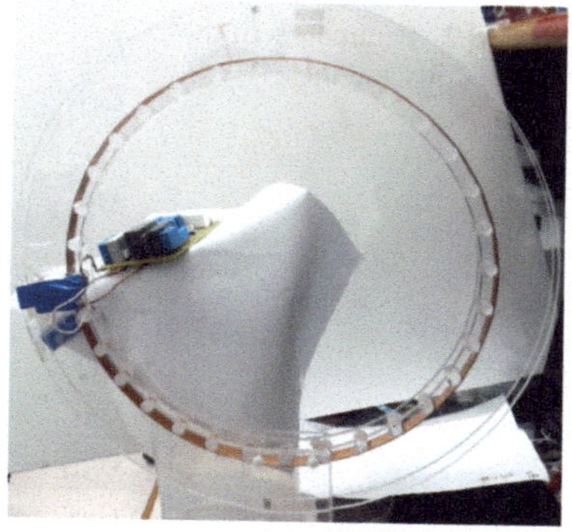

(b) Set-up of the secondary side

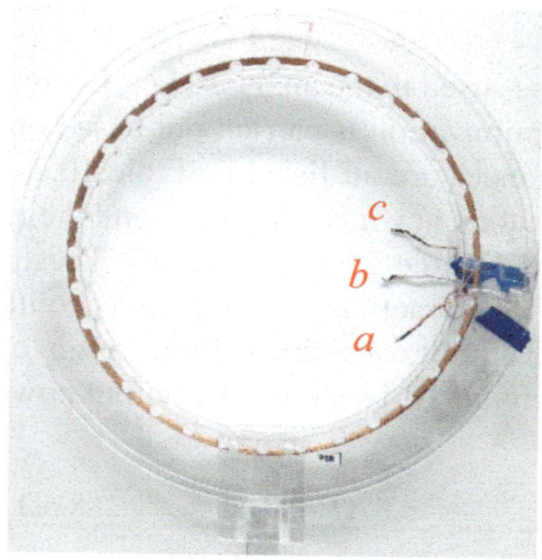


(c) Receiver coil splitting

**Fig. 10.16** Setup of the proposed system

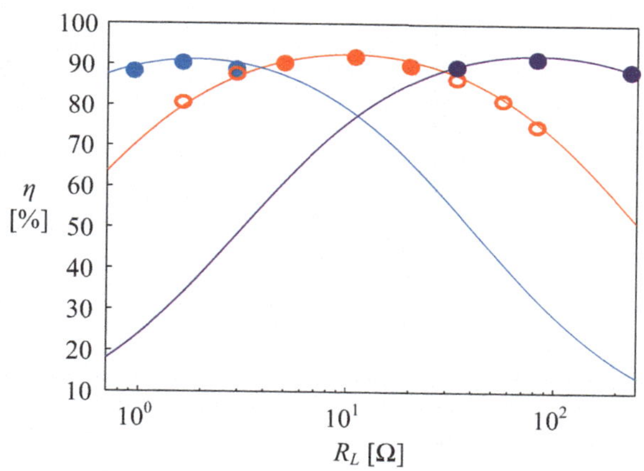


**Fig. 10.17** Measured efficiency of the proposed universal system (solid dots); measured efficiency of the two-coil system with load resistance not in the high-efficiency range (hollow dots); and the theoretical efficiency curves of three operation modes (real lines; blue: three-coil mode; red: two-coil mode; purple: two-coil with parallel capacitor mode)

**Table 10.2** Parameters of the splitting coils and the capacitors

| | |
|---|---|
| $L_{S1}$ | 77.22 μH |
| $L_{S2}$ | 1.55 μH |
| $M_{PS1}$ | 14.96 μH |
| $M_{PS2}$ | 1.38 μH |
| $M_{12}$ | 6.49 μH |
| $R_{S1}$ | 0.33 Ω |
| $R_{S2}$ | 0.04 Ω |
| $C_{S1}$ | 28.63 nF |
| $C_{S1P}$ | 27.27 nF |
| $C_{S2}$ | 924.5 nF |
| $C_{LP}$ | 51.30 nF |

little larger than the values used in the analysis, i.e., $L_{S2}$ is 1.55 μH (theoretical value 1.13 μH). Therefore, optimization is carried out based on the practical inductances and the optimal capacitances are all listed in Table 10.2. Figure 10.17 shows three sets of practical energy efficiency measurements and their corresponding theoretical efficiency curves under the three operating modes. These measurements agree well with the calculations. These practical measurements confirm the proposed idea that the reconfigurable WPT circuit can achieve high energy efficiency over a very wide load resistance range. Figure 10.18 shows the output power of the proposed system

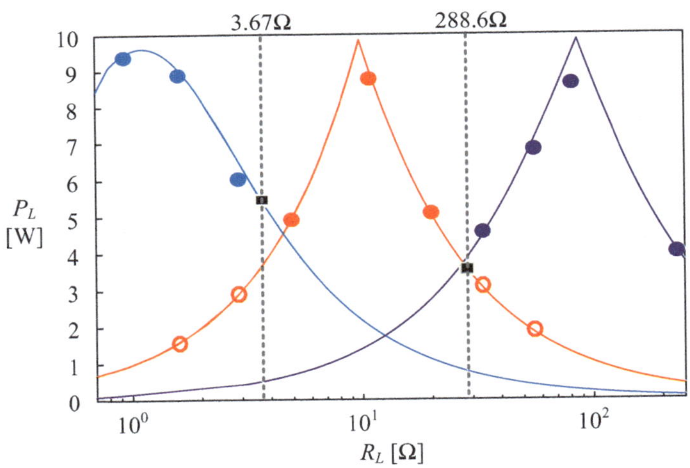


**Fig. 10.18** Measured output power of the proposed universal system with 1 A limitation in the coil current (solid dots); measured output power of the two-coil system with load resistance not in the high-efficiency range (hollow dots); and the theoretical output power curves of three operation modes (real lines; blue: three-coil mode; red: two-coil mode; purple: two-coil with parallel capacitor mode)

with a limitation of 1 A in the coil current. Again, the measurements totally verify the theoretical analysis.

## 10.5 Conclusion

By splitting the receiving winding, the optimal load resistance (at which high efficiency can be achieved) can be extended to a lower value. With the use of a parallel capacitor across the load, the optimal load resistance can be extended to a higher value. Taking advantages of these two complementary approaches, this study shows that using reconfigurable WPT systems can achieve high-efficiency operation over a very wide load range and simultaneously keeping the volt-amp ratings of the coils low. The proposal offers a new approach to achieve high efficiency in medium and high-power applications in which efficiency is of paramount importance.

The manufacturer's authorised representative in the EU is Springer Nature Customer Service Centre GmbH, Europaplatz 3, 69115 Heidelberg, Germany. If you have any concerns regarding our products, please contact ProductSafety@springernature.com

Printed and bound by CPI Group (UK) Ltd, Croydon, CR0 4YY
30/09/2026
02225574-0006